To Joe & R

in memory of Harry -

1992 -

With love from -

Katherine.

UNCERTAINTY, IN NATURE AND COMMUNICATION

UNCERTAINTY, IN NATURE AND COMMUNICATION

H. B. Rantzen

F.I.E.E., S.M.I.E.E.E., F.R.Met.S., B.Sc.(Eng.)

One time

Head of Transmission Design Dept.,
Standard Telephones & Cables Ltd.
Head of Lines Department, B.B.C.
Head of Designs Department, B.B.C.
Director of Telecommunication Services,
United Nations, New York

HUTCHINSON OF LONDON
Scientific and Technical Publications

HUTCHINSON & CO (*Publishers*) LTD
178–202 Great Portland Street, London W.1

London Melbourne Sydney
Auckland Bombay Toronto
Johannesburg New York

★

First published April 1968

This book has been set in Times, printed in Great Britain by Benham and Company Limited, Colchester, on Basingwerk Parchment paper, and bound by Wm. Brendon and Son Limited of Tiptree, Essex

09 086990 7

To Katherine, Esther, and Priscilla, without whom life would have been relatively uneventful

CONTENTS

LIST OF ILLUSTRATIONS

PREFACE

There are many reasons for writing a book. Some do it for financial gain, or out of a desire to entertain or instruct. In others, there is an almost irresistible urge to stimulate the imagination of those readers with whom they share a common cultural heritage, and to extend their intellectual experiences beyond the bounds to which they might otherwise have reached unaided. These are artists in whom the need for self-expression is paramount.

Most of the original work in this book was evolved during a decade in which I was professionally interested in the causes and statistics of the fading of radio waves, and in the atmospheric disturbances responsible for it. It seemed to me that these were phenomena which could not be explained by any existing 'science'. This new material has been collected and rearranged to form a single, consistent argument—for a reason which has recently grown so compelling that it is now as strong in me as is the need for self-expression in an artist.

The basic ideas were developed during three or four years of research into radio propagation immediately after my return to this country from the United States in 1952. During the next five years the accuracy of the general solution was improved, an analysis of the temperature distributions in the lower atmosphere was completed, and similar analyses were applied in turn to speech waves and gravitational perturbations. During this period I offered papers on these subjects to a number of learned societies, and collected in return a remarkable array of rejection slips. Those to whom I submitted my work were about equally divided between those who said they could not understand it, those who were prepared to dismiss agreement between it and measurements as 'coincidence', and those who thought that ideas which could not be proved rigorously were highly suspect. No 'Establishment' has ever been more firmly seated in authority than modern science in this machine age.

These 'failures to communicate' occurred piecemeal, that is to say, papers on the statistics, on meteorology, on radio propagation, were offered to appropriate learned societies and rejected one at a time. By the early 1960s I had become convinced that the vast majority of scientists were so deeply impressed by the success of their existing methods as to be regrettably complacent about the contradictions and fantasies on which some of their work is based.

Presumably in part because scientific axioms are so 'unnatural', an enormous gap has now opened between the scientist and the rest of humanity. There are, it seems, two worlds, one man-made and scientific, the other natural and human. Because anything made by man must be strongly influenced by his mental processes, which have evolved naturally, and because, nevertheless, science should be as objective as possible, it ought to be possible to build some sort of bridge connecting these two worlds, to construct a hypothesis which is in part coherent and subjective, and in part random and objective. I will be well satisfied if this book contributes in any way to this end.

I would like to express my thanks to Mr. J. W. Head, of the Research Department of the Engineering Division of the B.B.C., for advice which has led to an improved explanation of the normalised solution of the statistics.

London N.W.2 H.B.R.

PART I

UNCERTAINTY *per se*

All nature is but art unknown to thee,
All chance, direction which thou canst not see;
All discord, harmony not understood;
All partial evil, universal good;
And, spite of pride, in erring reason's spite,
One truth is clear, Whatever is, is right.

1

LAW AND DISORDER

The Law is the true embodiment
of everything that's excellent.
It has no kind of fault or flaw,
And I, my Lords, embody the Law.

Apologia

Once upon a time, at the birth of some new idea or scientific discovery, an Inquiring Mind first identified a few important causes of change, and then isolated them as far as possible, carefully arranging conditions so that the effects of all other causes of change could be neglected. He was thus able to formulate a *Natural Law* (which now, I hope, bears his name), expressing an approximate relationship between his measurements and the important causes of change identified by him, so that the many who followed could quickly and easily calculate what to expect for the future in similar circumstances. Within the next few centuries, very little in the evolution of man, but a most important part of his more recent history, there appeared scores of such natural laws, and these gave such widespread powers of controlling and anticipating events as to lead to a major industrial revolution. Men hastened to band themselves together into companies and in factories to design and fashion the machines, materials, and structures necessary to set up the special circumstances in which their laws would hold good. Vast changes appeared in the means of transport of men and goods, in the types of carefully selected materials and their uses, in environment particularly as it affects health, convenience, and comfort. Within two or three centuries there emerged a new modern world, civilised and scientific, so quickly as severely to tax the powers of expressed in rigorous mathematics, any new laws describing them must be relatively simple and include a sufficiently small number of symbols, or parameters, so that the result of theoretical calculations using them may be compared with actual measurements. There is no other way of justifying (or rejecting) them.

Since there is, as yet, no accepted symbolic form or mathematics dealing with the particular type of statistical distributions which will here be introduced for this purpose, much of the argument will inevitably be in clear English rather than in symbolic form. And since perturbations are always involved, most of the terms will be used with their ordinary dictionary or encyclopedia meanings, and not with the more precise meanings given them in technical dictionaries, in which it is very often assumed that the 'extraneous' disturbances of special interest here do not exist.

To those to whom the mathematical expression of a rigorous law represents 'absolute' truth, much of what follows may seem unscientific; to those who demand the utmost precision in all assumptions, it may seem vague and incomprehensible, even though changes in the real world about us never quite conform to rigorous laws, nor do we ever know all about them precisely. It is the hope of the author that the uncertainty in terms and laws which he has deliberately introduced will prove acceptable simply because it leads to a useful approximation to conditions in the natural world of which we are part. More precise methods of analysis surely have their part to play in the artificial world which today so many labour so successfully to create.

Uncertainty in mathematics

Any suggestion that mathematics, the very language of so much modern scientific thought, might in some way be inadequate, is one which very many will accept only with reluctance, if at all. Since it is a language, of course, it should ultimately be possible to extend its vocabulary to meet any needs. In the meanwhile, it is proposed here to point out some respects in which it may at present fall short.

In the practical applications of arithmetic to real objects, quantity is measured by number, the essential characteristic in any one number, and in the difference between one number and another, being not that there is a precisely uniform scale on which consistent measurement may be based, but that there is simply a count of a number of qualitative changes. If six apples are considered as a group, for example, it is most unlikely that they will together contain exactly six times as much of any 'apple' characteristic or quality as does one of them; the number 'six' is merely a record of the number of a particular type of statistical sub-distribution[1] of small changes of colour, of taste, of density, or what you will, that we have recognised in each one of the whole group. The larger changes from one of the sub-distributions to another are not abrupt; the edge of an apple is, obviously, not a precise point or area. Neither are the qualitative conditions within each apple necessarily all absolutely uniform; there is a range of fluctuation of the qualities, sub-distributions of their magnitudes, within any one of them. There is, however, at each transition, a change-over from the statistics within the apple via those of the dividing media, if any, to the similar statistics of the next apple. Numbers may only *theoretically* represent precise quantity and differ one from another by exact amounts. In any *real* application, the number of articles records only the number of statistical sub-distributions each identifying one article and separated in some way from the next. Even in an alleged continuous scale of quantity, measurement always involves the same necessity of points of change-over. There has to be a recognition of the beginning and end of any unit on the scale of measurement, some non-uniformity in it, before a quality can be assigned a quantity or number of such units. There is, in reality, always some variation of quality, and some uncertainty, therefore, within the unit or quantity measured, as well as between one unit and another. It is frequently claimed that the divisions or marks on a scale can be made 'as small as you like'. This is clearly untrue if there are uncontrollable perturbations both in what is measured and in the measuring scale. It is in general not possible, for example, to identify what is the length of a body with a precision greater than the mean free path of the particles of which the body and measuring scale are composed. For completeness, arithmetical quantity should always be supplemented by the uncertainty within and between the recognised sub-distributions or units involved in the measurement process.

Geometry is that branch of mathematics which deals with measurements in space, and the same limitation therefore applies

[1] A 'sub-distribution' is a small part of a whole distribution. A 'statistical sub-distribution' is one within which the relative frequencies of occurrence of small changes, when plotted as a graph, have a recognisable characteristic or shape, which, to a useful degree of approximation, may be expressed mathematically.

to it. One of its simplest laws relates the distance along the circumference of a circle to that across its centre. The circle is generally defined as a plain figure with a boundary which has no thickness; by definition, circles are therefore quite intangible, and can produce no effects on any of our senses. As far as we are concerned they just do not exist. As soon as we consider a real figure which approximates to a circle, the law relating distance along its perimeter to its overall size fails because of the width of the boundary and when the distance along the perimeter is so small that there is uncertainty about the direction of measurement. The geometric law is valid (approximately true) for practical cases only when the distance along the perimeter is very large compared with its thickness.

Finally, one of the most widely applied branches of mathematics in modern technology is the calculus, which is the science of infinitesimally small differences. The calculus may be used to derive important consequences from basic laws by considering the characteristics of indefinitely small changes in accord with such laws. In reality, of course, no basic law remains true for indefinitely small changes, and deductions which depend on the continued validity of a law no matter how small a change is made must, in the end, fail.

Some may by this time be asking why, if rigorous mathematics is so full of such pitfalls, science based on it has resulted in such spectacular advances. The fact is, of course, that progress in the modern world has become more and more a matter of the production of machines and refined materials, of controlled environment in which such limitations are deliberately, and often with the greatest difficulty, kept negligibly small. There exists, in fact, a whole range of devices—lubrication, tyres, water-jackets, insulation and the like—devoted exclusively to this end. But although the artificial environment of the modern world is becoming more and more of this type, we have still to adapt ourselves to a natural world the disturbances in which cannot always be so controlled, and it is just with this area of change that we shall mostly be concerned here. The essential feature of all rigorous analyses lies in the assumption of the consistent validity of one basic law at a time. The obvious fact in the natural world around us is that change of scale sooner or later inevitably involves a change of law, and often a basic change in the nature (and definition) of the parameters included in the law. Detailed study of circumstances which include natural perturbations involves a study of conditions in which changes in accord with at least two different laws must be considered together.

Uncertainty in statistics

The complete set of measured, or anticipated, performance figures, for a piece of equipment, or a complete system, over the full range of its possible adjustments and for all the various natural conditions in which it may have to operate, forms far too complex an assembly of data for direct use in design or planning. There are at least three possible 'shorthand' methods of expressing assemblies of data of this type:

(a) A rigorous expression involving a reasonably small number of important single-valued parameters[1] may be formulated, it being assumed that, for stated and useful ranges of adjustments and conditions, all other sources of change have negligible effects. Most of the so-called 'natural laws' and the vast body of science which surrounds them are examples of this. Such a use of rigorous mathematics is often highly successful in controlled conditions, but it is often equally unsuccessful over ranges of magnitude and time within which conditions are uncontrolled or turbulent. One rigorous law can never be true both over indefinitely large and indefinitely small changes and time-intervals, and the omission of reference to the ranges over which it is approximately correct may be very misleading.

(b) In some fields in which the important sources of change cannot all be identified, it may be helpful to assume a particular shape of statistical distribution to which, over useful ranges of time and magnitude, the frequencies of occurrence of changes closely conform. In such cases, values averaged over very large numbers of events, or over very

[1] A 'parameter' is any quantity entering into a mathematical expression or into the equation to a curve. It is usually defined as having a single value in each case, though it may, of course, have different values in different cases. The term 'single-valued parameter' has been introduced in contradistinction to a parameter which is not single-valued in each case, but is a statistical sub-distribution of values. The 'temperature' of a volume of gas, for example, does not precisely define the movements of all the gas molecules in that volume. It has not a single value at any one moment. It can often more usefully be regarded as a sub-distribution of 'temperature disturbances' within smaller volumes (sub-volumes) of the gas.

long times and very large magnitude ranges can, with sufficient accuracy, be expressed in rigorous mathematics, sometimes simply because they are substantially constant. In addition, although it may not be possible to account for individual occurrences, the 'shape' of the statistical distribution about its mean, the 'probability' of each change, can also be defined using rigorous laws and single-valued parameters with acceptable accuracy. This method of concentrating data has, in the past, been used with limited success is such fields as climatology, biology, sociology, demography, etc. Since particular events are not identified by this procedure, such statistics may be said to be *singly indeterminate*.

(c) It is possible to extend the method of (b) above, even to fields in which long-term averages cannot be usefully expressed by single-valued parameters, by assuming that these averages, as well as individual 'events' within the field, are themselves statistical sub-distributions each of known shape and each defined by a reasonably small number of single-valued parameters. Very little practical use has so far been made of this type of approach, although general advice covering identifiable sections of such fields of sub-distributions has been given by Batchelor, Shannon, Rice, and others. Since in these cases each change within the total field is singly indeterminate in its associated sub-distribution, and since the hypothesis deals only with the frequencies of occurrence of sub-distributions which are themselves also singly indeterminate in the whole field, there is, in the complete distribution, a *double indeterminacy*.

In doubly indeterminate distributions, there are two different statistical characteristics (each the result of what will be called its own 'causal set'), one applying to the distributions of sub-distributions, and the other applying within each sub-distribution. These two statistical characteristics and causal sets, *and only these two*, are regarded as responsible for, and 'explaining', all changes within the range of magnitude and/or the time-interval under consideration. Changes in the neighbourhood of a particular magnitude not due to one of the causal sets must therefore be due to the other: that is to say, the total probabilities cannot be due only to either one of the two statistical characteristics; each characteristic must be indeterminate to the extent of the probabilities of similar changes in the other, and vice versa. In doubly indeterminate distributions, small variate[1] changes therefore occur with frequencies mostly in accord with the 'sub-distribution' statistical law, but these frequencies of occurrence are, to some extent, indeterminate as a result of the statistical law and causal set of the large variate changes which embrace many sub-distributions. Frequencies of occurrence of the latter larger changes accord mostly with their own statistical characteristic, but their total probabilities are to some extent indeterminate as a result of the small perturbations within each sub-distribution. The whole statistical field may of course consist of changes in magnitude at the same place occurring one after the other in time, or of simultaneous magnitude changes at different places. In some cases of particular interest, both of these space and time distributions will be considered together.

One obvious feature of such a doubly indeterminate distribution is that there must be some intermediate events of which the total probability comes equally from both causal sets. In these cases, the probabilities are equal to their own indeterminacies, and it is impossible to tell, from measurement or observation, whether the changes lie within the sub-distribution or are the result of differences between one sub-distribution as a whole and another. The minimum value of the probability at which there is such a 'change-over' from the statistical law of the large-scale variates to that of the perturbations within sub-distributions will be here referred to as the 'Basic Uncertainty' in the distribution of the large-scale variations. If, as will be generally the case here, the whole distribution is divided into sub-distributions each with this particular 'Basic Uncertainty' probability, then the latter will form, and may be referred to as, the 'Quantum of Probability' in the analysis. As a practical example, consider the case of a communication circuit in which there are both signal changes and noise perturbations. Signal differs from noise simply because changes in it conform to some recognisable law or coherent rule which we have previously learnt as part of a code or language. In general, the smaller the signal change that occurs, the more difficult it may be to recognise it as part of the code or language. However that may be, there will ultimately be a change so small that, considered alone and out of context, it may just as probably be noise or perturbation as signal change. A small change considered alone conveys no significance or information when its probability is

[1] A 'variate' is any quantity which varies.

such that it might just as likely be noise as signal, and we are then left completely uncertain as to which it is; it is the basic uncertainty in the signal, the quantum of probability below which changes cannot statistically be recognised as signal.

In addition to the cases to which methods (a), (b), and (c) above apply, there are, of course, many fields in which disturbances are so random that frequencies of occurrence neither of individual events nor of sub-distributions can accurately be specified, and indeed individual events and sub-distributions cannot clearly be separated one from another. In such cases all three of the above methods of analysis fail completely; most long-term weather and fall-out predictions, political developments, and most evolutionary and economic changes, at present come into this category.

It should be observed that all three types of analyses (a), (b), and (c) are approximations in which, with varying degrees of success, the effects of very large numbers of unknown causes of change are considered negligible. Whether this is really justifiable must ultimately be a matter of point of view. The lower atmosphere, for example, is highly turbulent to human beings living in it, but it might appear to be a very stable and constant medium to a 'large-scale' observer on another planet interested only in its overall characteristics and changes in them from year to year. It may also be noted that the success of the statistical methods of (b) and particularly of (c) above depends very much on the way in which data are prepared for use. They involve recognition of hypothetical statistical sub-distributions and distributions the shapes of which are simple enough to be defined mathematically. This cannot usually be done merely by dividing the field into convenient time and/or magnitude ranges. Fields of change in turbulent media can be analysed only if sub-distributions are chosen so that they *always*, with sufficient accuracy, conform to the particular statistical shapes of characteristics on which the analysis is based. In general, the adoption of *any* constant variate range and/or time-interval does not lead to a constant shape of sub-distribution, and is, therefore, not a guarantee of consistent results.

Analysis is, by definition, the resolution of a complex into simpler elements, and it is successful when these elements can be recognised as conforming approximately to some simple law or as part of some recognisable order, and their synthesis leads to an acceptable approximation to the original complex. In analyses of natural changes, the law assumed to be responsible for each event or change applies to the complex, and each parameter ought rather to be an assembly or sub-distribution of perturbations, and is far too complex for complete specification by any one single-valued parameter. The natural world is basically statistical, and detailed analyses of conditions in it always lead to multi-indeterminate statistics. The doubly indeterminate distributions of (c) above constitute merely a second order of approximation to the truth, useful perhaps when rigorous analyses are not detailed enough, or when natural perturbations are so large that they constitute turbulence.

It is, of course, an essential element of rational thought that there should be in it at least some, perhaps a maximum, of orderliness or coherence—some adherence to a law or purpose simple enough for us to recognise. Indeed an *organic* body is as a rule recognised as alive only if its behaviour is such that parts of it are free to move randomly, yet do so coherently. On the other hand, it follows from the second law of thermodynamics that any physical system in the natural, *inorganic* world, left to itself, tends towards a condition of minimum coherence or maximum randomness. There are always many more possible complex (incoherent) states than there are possible relatively simple (coherent) states—without interference from any living agency, therefore, natural changes tend toward incoherence. They tend toward the condition often described as having maximum 'entropy'—entropy being a measure of the uncertainty in our knowledge.

Understanding of changes in man's natural surroundings would seem therefore to involve two causal sets. One of these, arising perhaps from limitations in his mental capacity, reflects the basic need for simplification in his thought processes, and should accordingly be coherent. The other expresses the ultimate relative complexity of his natural environment, and should clearly be incoherent or random therefore. The simplest possible distribution arising from two such causal sets will here be called the Neogaussian Distribution. We propose now to discuss this in some detail.

2

THE NORMALISED NEOGAUSSIAN SOLUTION

> Great fleas have little fleas upon their backs to bite 'em,
> And little fleas have lesser fleas, and so *ad infinitum.*

So far, the great majority of 'natural laws' have been formulated in rigorous mathematics. Although the number of the others is increasing, there are still so few of them flexible enough to admit possible perturbations that, once the need to do so is appreciated, it becomes a comparatively simple matter to select the two causal sets and statistical characteristics appropriate to a doubly indeterminate distribution intended to describe fields of natural changes, which always include perturbations.

First of all, however, some explanation of terms seems desirable. The 'Distribution' is the *whole* continuous field of changes under consideration, not just isolated sections of it. Small parts of this distribution will be referred to sometime as 'Events', more often as 'Sub-distributions'; they contain the perturbations. The analysis will be directed towards identifying events, or sub-distributions of 'perturbations' or 'sub-events', each by its consistent and recognisable statistical characteristic 'Pattern' or 'Shape' about its mean; the mean of a sub-distribution of perturbations might appear as one value of a single-valued parameter in analyses which neglect the perturbations which, however small they may be, always exist about it. This pattern identification or recognition process is basic to all human sensibilities and techniques, and its omission from rigorous analyses is perhaps the main reason for the inadequacy of the latter in many turbulence and perturbation problems. We accept a wheel as an entity in spite of textural peculiarities within it—we have learnt by experience the wheel 'pattern', its round shape and axis of symmetry about which it can rotate, etc.

The large-scale distribution of sub-distributions will be attributed *mainly* to a 'Coherent Causal Set': the small-scale changes within each sub-distribution will be regarded as *mainly* the result of an 'Incoherent' or 'Random Causal Set'; just as a cause produces an effect, so does a causal set produce the range of effects which constitutes a statistical distribution or sub-distribution. Small changes which come from the incoherent or random causal set, and groups of sub-events which, as far as can be seen, come mostly from this causal set, will generally be referred to as 'Perturbations'. Similar small (and larger) changes which come from, or mostly from, the coherent causal set will generally be called 'Fluctuations'. The statistical distribution which would result from either of the causal sets alone can, it will be assumed, be defined rigorously. This involves the unrealistic assumption that each sub-event or perturbation is a single-valued change—that there are no sub-sub-events or sub-perturbations within it. It leads to a second order of approximation to the truth, just as the common neglect of perturbations, and the assumption of a single rigorous law for the large-scale distribution alone, might lead to a first-order approximation to the facts.

The average of an increasingly large doubly indeterminate field of sub-events may *tend* towards a precise mean, the latter being one of the rigorous parameters assumed for the large-scale distribution. It can, however, never with certainty reach one precise value, however large the total distribution considered, since there will always be the 'perturbation' indeterminacy of at least one sub-distribution, within which sub-events occur with frequencies which accord with quite a different law. This ultimate indeterminacy may be very small, but we are here particularly concerned with cases in which it cannot be neglected, and with the limitations which its presence sooner or later always introduces. Consider for example the throwing of a perfectly balanced die in ideal conditions, there being the numbers one to six on its sides. No matter how often it is thrown, the average of a 'very large number' of throws can never with certainty be $(1+2+3+4+5+6)/6=3\frac{1}{2}$; for if the number of throws is raised to the point at which the difference (if any) between the experimentally determined average and $3\frac{1}{2}$ is considered to be completely negligible, there will still be the uncertainty of at least one more throw, and of the associated 'equiprobable' sub-distribution one to six in any complete distribution under review, an uncertainty not removed by considering any indefinitely large number of throws. All we can do is to reduce the expected effects of one more throw on the average of *all* the preceding throws by increasing the number of throws. We can never eliminate it altogether, since it would be absurd to postulate so many throws that at least one more could not then be made. The average of the complete distribution will always finally have the incertainty of at least one sub-distribution, however small a part of it that may be. As soon as we admit the presence of a causal set producing perturbations within sub-distributions (the reasons why throws differ one from another), probabilities in the large-scale distributions and all the parameters defining them become, to some extent, indeterminate.

Similarly, as we consider smaller and smaller fields, the rigorous assumptions defining the large-scale distribution must fail at a field of one sub-distribution. We have assumed a different statistical shape within the sub-distribution, one appropriate to the perturbations within it, and, as a result, nowhere in the large-scale distribution can frequencies of occurrence be determined more closely than to within the probability of one sub-distribution. This limiting sub-distribution has thus the 'Quantum of Probability' in the statistical characteristic of the large-scale distribution, in which it is the 'Basic Uncertainty'.

The presence of perturbations in this way imposes both upper and lower limits on the size of field which can be defined by one single rigorous law for the whole distribution. Outside these limits, the coherent law assumed for the large-scale distribution is no longer valid, there being a 'Change-over' either to the statistical law assumed for the sub-events or perturbations within the sub-distribution, or to some other law expressing changes in the mean of the whole field during still longer times and/or over quite different magnitude ranges. With this in mind, the 'Probability' of a particular part of a doubly indeterminate distribution will be defined as the relative frequency of occurrence of the sub-distributions within that particular part, with a minimum indeterminacy or 'Basic Uncertainty' equal to the probability P(R) of one sub-distribution, *the total number of sub-distributions in the whole distribution being* $1/P(R)$. The sub-distribution is, of course, one within which sub-events are distributed according to the other (small-scale) statistical law implied in the term 'doubly indeterminate'. This definition leads to a natural scale of probability in integral values of that of the sub-distribution or 'Quantum of Probability' P(R), all such values being indeterminate by at least P(R).

It has already been pointed out that although, for subjective reasons, analyses have always to be relatively simple and coherent, distributions of natural changes, objectively, are always much more complex, and tend towards their most probable state of minimum coherence. Clearly then it is reasonable to assume that the causal set mainly responsible for the perturbations within each sub-distribution is one which *alone* would produce what is generally described mathematically as a 'Rectangular' distribution. This is the one in which the tendency to a mean is the absolute minimum, and in which all sub-events or groups of sub-events within the same deviate range are equally likely. But since there are actually two causal sets, the sub-distribution statistics cannot be due wholly to this incoherent causal set. We shall describe the *approximately* rectangular sub-distribution which results in these circumstances by the term 'Equiprobable', which, although it has appeared in the literature from time to time, seems not yet to have graduated to the better-known dictionaries. The equiprobable sub-distributions, in a doubly indeterminate field, which result mainly from the incoherent causal set, are ones in which all sub-events over the same range of deviation are as nearly equally probable as the indeterminacy which results from a relatively small number of fluctuations from another, the coherent causal set, will permit. This statistical shape of sub-distribution is the one objectively to be expected in natural perturbations from the second law of thermo-dynamics; it expresses the *tendency* to a state of maximum entropy and minimum coherence in small parts of a doubly indeterminate field.

The other causal set, the one responsible for the small number of coherent changes in these equiprobable events or sub-distributions, follows from the very nature of any successful quantitative analysis in which there has to be some coherence and a tendency, therefore, to a mean of the fluctuations over a large magnitude range and/or long time-interval. If very many similarly 'shaped' sub-distributions tend toward a mean, it is clear from the Central Limit Theorem that the simplest possible Coherent Causal Set responsible for their statistical distribution should be one which, considered *alone* (without perturbations), would result in a Normal or Gaussian distribution of events or sub-distributions. The latter, of course, covers a range of magnitude and/or a time-interval which are large compared with those of the sub-events or perturbations coming from the incoherent causal set.

The combined result of both these causal sets will then be the Neogaussian Distribution, which is simply an attempt to add the effects of perturbations to the Normal Distribution. It expresses the limitations imposed on the validity of the exact Gaussian 'law' when it is admitted that there are, in fact, two simultaneous causal sets and laws, one coherent (in accord with the Central Limit Theorem), mainly determining the large-scale effects, the other (in accord with the Second Law of Thermodynamics) incoherent or equiprobable, in the small perturbations. It constitutes a second order of approximation to natural distributions of probably the simplest and commonest type, since the two contributing causal sets have hypothetical statistical laws which are those of the Gaussian distribution, towards which similar sub-distributions recognisable by coherent human beings always tend, and the equiprobable, or random distribution of maximum entropy, characteristic of completely incoherent perturbations. In the Neogaussian distribution there are these two, and only these two, causal sets responsible for the frequencies of occurrence of changes. They are not completely independent in the total distribution, since what is not due to one of them must be the result of the other. As has been already pointed out, probabilities from one of them are indeterminate to the extent of the probabilities of similar changes due to the other.

Even with these simplest possible theoretical assumptions about the statistical characteristics of the two contributing causal sets, the number of unknowns involved in their combined distribution would be too great for any general solution without two additional simplifications:

(a) The first of these follow from the 'pattern recognition' process basic to all analysis, which is essentially one of identifying the predominant coherent changes in larger parts of the field down to the smallest detectable coherent change, which is that in the equiprobable sub-distribution in which all similar coherent and incoherent changes are equally probable and therefore indistinguishable. The point of maximum analytical sensitivity will obviously be reached when the probabilities of small parts of the coherent distribution fall to the same value as that of the equal incoherent perturbations in the same equiprobable sub-distributions, and therefore to that of their own uncertainty. If each sub-distribution has this quantum of probability, then clearly there will be a reduction in the total number of parameters to be evaluated, and the distribution can be considered in sub-distribution size ranges, one above and one below this size, over each of which the assumption that the statistical parameters are all single-valued, and that there is only one causal set and law valid at a time, is reasonably accurate. We can use a footrule to measure length down to the uncertainty resulting from the width of its scale marks. Above this uncertainty, we need consider only the law of the scale as a whole; below it, we might achieve a closer approximation by considering the statistical characteristics of changes within small parts of the marks.

Firstly, then, the whole distribution must be analysed in small equiprobable (approximately rectangular) sub-distributions or events of constant probability, the Neogaussian 'Probability Quantum', as it will be called, within which any small change is equally likely to come from either causal set. This Neogaussian Probability Quantum is the basic uncertainty of the whole distribution within which the effects of the two causal sets are inextricably mixed. Perturbations in sub-distributions of lower probability are likely to be the result mainly of the incoherent causal set, and within them the effect of the coherent causal set appears only as a small indeterminacy in their means. Parts of the distribution of appreciably larger probability than the Neogaussian Probability Quantum will have the statistical characteristics of the Neogaussian distribution, which, over this range, tend toward those of the Gaussian distribution; the equiprobable components within their fluctuations produce only a small 'perturbation' indeterminacy of mean. The Neogaussian Probability Quantum itself thus marks the change-over between two domains covering different orders of magnitude, within each of which one of the causal sets is predominant, and outside each of which the law of the other is the more appropriate. It is the point at which the coherence of the Gaussian causal set is just lost in perturbations; in communication parlance, it is the point at which 'signal' falls to 'noise'.

(b) The other simplification is the familiar device of normalisation. In both causal set characteristics, probabilities depend on deviation. The frequency of occurrence of small changes from the coherent (Gaussian) causal set falls rapidly with increasing deviation,[1] while that of similar small changes from the incoherent (Equiprobable) causal set remains the same for all deviate values. Normalisation can be applied to reduce the total number of independent variables by defining as 1·0 the deviations in the neighbourhood of which similar small changes come with equal frequency from both causal sets. Of course, the 'real' value of this deviation of 1·0 in practical cases must be found from the physics of the problem or by actual measurement. All the other parameters of the two distributions are expressed in terms of it. Its actual value is determined by the relative numbers of similar contributions from the two causal sets, the relative importance of the perturbations in the whole doubly indeterminate distribution.

With these two added simplifications, a general solution of the Normalised Neogaussian Distribution may now be derived. There will be, in the neighbourhood of any deviate in the Gaussian Distribution, a small range of coherent fluctuation the probability of which is just equal to that of similar but incoherent perturbations within the same event or sub-distribution. When this deviate in the Gaussian Distribution is 1·0, the range over which associated equiprobable incoherent sub-events can extend is, by definition, also 1·0, that is to say, from 0·5 to 1·5. This follows directly from the normalising; it is the way the statistical characteristics of the two causal sets have been adjusted one to the other so as to yield one general solution for their combined distribution. In the range from 0·5 to 1·0, incoherent sub-events will be heavily outnumbered by coherent contributions—the frequencies of ocurrence in the Gaussian (and in the Neogaussian) characteristics rise sharply with decreasing deviation. This part of the deviate range of the Neogaussian Distribution is certainly not equiprobable. From 1·0 to 1·5, on the other hand, the number of coherent contributions would continue to fall rapidly, and they are therefore heavily outnumbered by the incoherent perturbations, the 'probability densities' of which are constant at all deviates. This part of the range is, therefore, to a second order of approximation, equiprobable. It will be noted that we are here neglecting third-

[1] The 'Deviation' at any point (deviate) is the difference between the variate value at that point and at the mean.

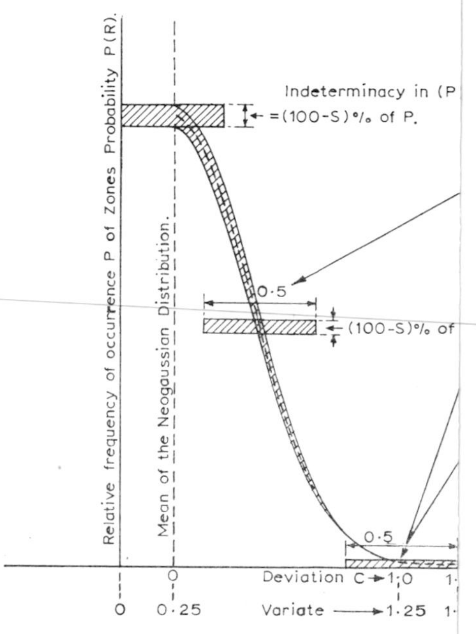

The normalised solution for a maximum
in terms of the sub-distribution of Prob
Indeterminacy (100−S) = 1·095%
Percentage Coherence S = 98·905%
Standard Deviation $\sigma = 0{\cdot}4363$
Basic Uncertainty $P(R) = \sigma^2 \times (100-S)/1$
(Quantum of Probability)
Maximum Probability up to middle of la
sub-distribution = 0·9845
Mean of the whole distribution = 0·25.

Fig. 1. The Normalised Neogaussi

order effects during the actual change-over, and within the relatively very small range of deviation of the coherent fluctuations within any one sub-distribution. The equiprobable sub-distribution in the perturbations associated with a coherent deviation of 1·0 thus extends from 1·0 to 1·5, with a mean close to 1·25. Since, by definition, all the equiprobable sub-distributions into which we are dividing the whole distribution have the same probability, they will all extend over this same range of deviation, up to 0·5 further from the Gaussian mean (the point of zero coherent deviation); only the small deviate range of the coherent fluctuations in each sub-distribution varies with average deviation, falling markedly with decreasing average deviation. If all the equiprobable sub-distributions are centred on mean deviates 0·25 more than the average Gaussian deviation of their coherent fluctuations, the effect of the incoherent perturbations must be to add 0·25 to the mean of the whole normalised Neogaussian Distribution. By adding the perturbations, we have increased the number of sub-events in the whole distribution; this inevitably increases its mean. The sub-distribution size and normalising have been 'adjusted' so that this change of mean is 0·25 (see Figure 1).

This mean change of 0·25 is one of three parameters which define the Normalised Neogaussian distribution. Each of the sub-distributions has a probability of one Neogaussian Probability Quantum, with equal uncertainty (probability in the incoherent statistical characteristic) as a result of the perturbations within it. Each includes a relatively small range of coherent fluctuations at the maximum deviation of 1·0 from the mean of 0·25 (i.e. at a variate of 1·25), this range decreasing rapidly with lessening deviation values. The density of the sub-distributions of constant probability rises sharply as the coherent deviation falls. The associated incoherent perturbations in the equiprobable sub-distributions extend over a constant deviate range of ±0·25 about these small coherent ranges, for the whole range of coherent deviation from 0 to 1. Since all the sub-distributions have the same probability, the number of them in any constant small coherent deviate range, their density, must follow the Gaussian characteristic curve, and since the indeterminacy (the probability of the incoherent components in their own statistical distribution) in each sub-distribution is constant, the total 'Indeterminacy' in the coherent probability densities in the neighbourhood of any point in the Neogaussian curve must be a constant fraction or percentage of the Gaussian probability density at that point (see

Figure 1). The second parameter is this Indeterminacy, which we will denote by (100-S)/100 or (100-S)%. The probability density in the neighbourhood of any point in the Neogaussian characteristic is thus S% coherent with a (100-S)% indeterminacy range about it. The average statistical shape of the whole distribution of sub-distributions is Gaussian, from the mean of 0·25 to the maximum coherent deviation 1·0 beyond this mean, at a variate of 1·25. The third and last parameter defining the whole distribution is obviously the Standard Deviation σ, or Variance σ^2, of the sub-distributions within this range, each of which has a probability (and an uncertainty) of one Neogaussian Quantum of Probability.

The relationships which can be used most easily to calculate (100-S)% and σ follow also from the assumption that there are two, and only two, causal sets, so that sub-events not due to one of them must come from the other. If the total probability in the combined distribution and 'under' the rigorous characteristic curve assumed for each of the two causal sets alone is 1·0, the overall indeterminacy in the total probability up to any deviate in one of them must be due to the other causal set, and it must be equal, therefore, to the probability of all sub-events from the same causal set under the 'tail' of the curve *beyond* that deviate. This important conclusion follows directly from the 'second-order approximation' assumption that there are two, and only two, causal sets. Hence, if all probabilities up to a deviate of 1·0 are (100-S)% indeterminate, the probability under the 'tail' of the Guassian curve beyond a deviate of 1·0 must also be (100-S)/100. Symbolically, if C is any deviate

$$\frac{1}{\sigma\sqrt{2\pi}}\int_{1\cdot 0}^{\infty} e^{-C^2/2\sigma^2}\, d\,(C) = (100\text{-}S)/100 \qquad (1)$$

From the other point of view, all the sub-distributions are equiprobable, and have an indeterminacy of mean which comes from the fluctuations from the coherent causal set. Let ΔC be the small coherent deviate range (indeterminacy of mean) in the sub-distributions at a mean coherent deviate of C. Then the ratio of the probability densities within these sub-distributions, i.e. at deviates $(C - \Delta C/2)$ and $(C + \Delta C/2)$, must be

$$\exp\left[-\left(C-\frac{\Delta C}{2}\right)^2/2\sigma^2\right] \Big/ \exp\left[-\left(C+\frac{\Delta C}{2}\right)^2/2\sigma^2\right]$$

$$\approx \exp\left[-\frac{1}{2\sigma^2}(C^2 - C\Delta C - C^2 - C\Delta C)\right]$$

$$\approx \quad 1 - \frac{C\Delta C}{\sigma^2}:$$

but since the probabilities at all points in the curve are only S% coherent, this is also S/100. Hence

$$1 - \frac{C\Delta C}{\sigma^2} = S/100 = 1 - (100\text{-}S)/100$$

so that

$$C\Delta C = \sigma^2\,(100\text{-}S)/100.$$

At the maximum coherent deviate, C=1·0, so that ΔC equals σ^2 (100-S)/100. At this, the maximum coherent deviation of 1·0, there is a single end sub-distribution in which the coherent fluctuations cover the deviate range ΔC, equally probable incoherent perturbations extend up to a deviate of 1·25, and the whole sub-distribution has a probability (and uncertainty) of one Neogaussian Quantum of Probability. In this sub-distribution, the coherent and the incoherent perturbations, the latter covering a total range of 1·0 about the coherent deviate of 1·0 in their *whole* distribution, are equally probable and indistinguishable one from the other. ΔC at the coherent deviate of 1·0 is therefore numerically equal also to the probability of the incoherent perturbations, and to the Neogaussian Quantum of probability, and is the probability of all the other equiprobable sub-distributions into which we are analysing the whole distribution. Hence it is also equal to the Basic Uncertainty (P(R)), and the probability beyond the end sub-distribution which extends up to a deviate of 1·25, so that:

$$P(R) = \sigma^2 \times (100\text{-}S)/100 - \frac{1}{\sigma\sqrt{2\pi}} \int_{1\cdot 25}^{\infty} e^{-C^2/2\sigma^2}\, d(C) \qquad (2)$$

Expressions (1) and (2) above can be used to obtain S and σ separately. Perhaps the simplest way of doing this is to take advantage of the good work of able friends across the Atlantic, and use the fifteen-figure tables of normal probability functions prepared by the National Bureau of Standards in their Applied Mathematics Series—23. This leads to

$$\sigma = 0{\cdot}436281$$

$$(100\text{-S}) = 1{\cdot}09499\,\%$$

$$P(R) = \sigma^2 \times (100\text{-S})/100 = 0{\cdot}00208422$$

to six significant figures.

It is, of course, quite possible to obtain this solution to any number of figures that may be desired. There is, however, in the fundamental assumptions upon which it is based, a basic uncertainty of one-fifth of 1% approximately (P(R)), and furthermore there is the unrealistic assumption that there are no sub-perturbations within each perturbation. Figures that might reasonably be used in practical applications are:

$$\sigma = 0{\cdot}4363$$

$$(100\text{-S}) = 1{\cdot}095\,\%$$

P(R), the Quantum of
Neogaussian Probability $= 0{\cdot}002084$
These are correct to four significant figures.

In view of the great importance of this normalised Neogaussian solution in all that follows, it may be helpful briefly to retrace the more important steps in the argument leading to it, and as far as possible to attach physical significance to the main assumptions in the underlying hypothesis.

The statistical characteristics of the two contributing causal sets are perhaps familiar enough to need no further comment. The effects of the random causal set may be perturbations in a volume of gas all at a single temperature—or disturbances in orbit which result when a nearly spherical satellite spins on its own axis as well as rotates round a nucleus or sun—or the radiation coming from parts of a wave so small that they should rather be thought of as isotropic point sources along the wave-front. Such perturbations are often isotropic as well as random. The fluctuations from the coherent causal set (these often have a particular 'direction') become statistically indistinguishable from the random perturbations, and can therefore form recognisable equiprobable sub-distributions with them, whenever we consider so small a part of the coherent distribution that its probability is no greater than that of an equal perturbation within the same sub-distribution. A single equiprobable sub-distribution contains no identifiable coherence therefore. Coherence appears simply as correlation between a small change in one sub-distribution and a fluctuation in the next, the coherent causal set being defined as one in which there is some 'tendency to a mean', and therefore some correlation between its effects whenever they occur. The limiting sub-distribution in which the fluctuations are indistinguishable from perturbations can thus be defined only by its probability in the whole distribution. In part-distributions more probable than this limiting sub-distribution, which has what we have called the 'quantum' of probability, the coherence may build up to form temperature differences or gradients in the vertical plane, or recognisable parts of a circular or elliptical orbit, or of a coherent wave-form. In a communication circuit in which signal currents are flowing, sufficiently small changes of current cannot be distinguished from noise; a current change may be recognised as 'signal' whenever its probability is such that it can no longer be dismissed as random noise.

We start then by dividing up the whole doubly indeterminate distribution into equiprobable sub-distributions each of which has the quantum of probability. But if we consider the distribution as a whole, we can associate with any one fluctuation in any single sub-distribution a finite range of perturbations about it. As an example, the spins of satellites (which are never truly spherical) will introduce perturbations into their orbits, the perturbations extending over a finite range during each spin. In the hypothesis we have assumed that the perturbations are completely random, that is to say, there is in them no tendency to a mean, and that they extend over a range normalised to 1·0. There are a number of simple consequences of this:

(a) If all the sub-distributions have the same probability and the coherent fluctuations in them follow a Gaussian law, the range of fluctuation within any one sub-distribution must be much less near the mean of the distribution than toward the tail. If in any one sub-distribution we associate equal fluctuations and perturbations together, the probability density of the sub-distributions, as of the fluctuations and perturbations within them, must also follow a Gaussian law, and be much greater near the mean of the whole distribution than at the tail. Both causal sets embrace a total probability of 1·0, and it therefore follows that if the difference in their statistical characteristics is attributed to the density of the sub-distributions, the deviate range of both characteristics must be the same. The normalisation of the maximum deviate to 1·0 applies to the statistical characteristics of both causal sets therefore.

(b) The sub-distributions with their fluctuations and perturbations are typical of turbulence in which we can identify particular occurrences only through some patterning or coherence in them. The Neogaussian distribution is a distribution of recognisable equiprobable sub-distributions (Rayleigh distributions) in which these patterns are recognisable down to a threshold set by the perturbations, which provide both the background to, and the irrelevant detail within, each pattern. Pattern recognition based on some coherence is essential in analyses of turbulence, in the recognition of waves, and in the last resort in all other physical measurements. We have to be able to identify what it is we are measuring before we can measure it. The range of the perturbations (1·0) about any fluctuation at a deviate C along the Gaussian characteristic must extend, at deviates less than C, into parts of the distribution where there are relatively much larger densities of sub-distribution, these building up to form the coherent part (to the left of the shaded curve area of Figure 1) of what is effectively a Gaussian distribution of *nearly* coherent values of the variate. In this and all that follows, there is of course the basic uncertainty always present. From C to (C+0·5), however, the perturbations associated with the sub-distributions at C ((100-S)% of the Gaussian probability at C) build up to substantially a Rayleigh distribution over a constant range ±0·25 about a median value of (C+0·25), the sub-distributions over this C to (C+0·5) range being everywhere and at all times much more numerous than those associated with any larger coherent deviation. The whole shaded curved area in Figure 1 beyond any deviate C represents a probability much smaller than that of the equiprobable sub-distributions associated with the coherent deviate C.

The whole distribution can in this way be analysed as a succession of recognisable patterns (equiprobable (Rayleigh) sub-distributions) adding to a constant range ±0·25 centred on variate values 0·25 greater than would be expected from the frequencies of occurrence of the coherent fluctuations in the Gaussian distribution. In other words, the point of zero coherent deviation in the Gaussian distribution occurs at a variate value of 0·25 less than the mid-point of the recognisable equiprobable sub-distributions at zero deviation in the Neogaussian distribution. Over the range of coherent deviation from 0 to 1, the density of the sub-distributions falls from its maximum at the mean until there is only one sub-distribution at a deviate of 1·0. For this to be the case, the whole distribution must of course be divided into parts which extend only over the amplitude range and/or the time-interval of the single sub-distribution at a deviate of 1·0, which is therefore the maximum coherent deviation in the whole distribution. If the whole distribution is considered in parts which cover larger amplitude ranges or last for longer time-intervals than this, there will be more than one sub-distribution in each 'sample', and some of the information otherwise available in the data will be lost. In quanta each P(R) of the whole, only the basic uncertainty P(R) is lost.

(c) The normalised solution follows from three simple and fairly obvious facts:

(1) Since the whole distribution has a constant percentage ((100-S)%) of fluctuations in all its equiprobable sub-distributions over the whole coherent deviate range from 0 to 1, the indeterminacy at any point (departure from a rectangular statistical characteristic) due to the coherent fluctuations in the equiprobable sub-distributions must be (100-S)% of the probability at that point, and the probability under the Gaussian curve beyond the end sub-distribution must be the product of this indeterminacy and the variance. The variance of course is the parameter of the Gaussian distribution, its mean squared deviation, which allows for the uneven probability densities.

(2) The probability beyond the end sub-distribution must be the basic uncertainty of the whole distribution and therefore equal to the probability of one sub-distribution (equation 2).

(3) If the indeterminacy over the whole range of coherent deviation is a constant percentage (100-S)% of the Gaussian probability, the probability beyond the maximum coherent deviate must also be this same constant percentage (equation 1).

It is a comparatively simple matter to check the normalised solution using the fifteen-figure statistical tables to which reference was made above. The probability under the single tail of the Gaussian distribution beyond $1/\sigma$ standard deviations should be (100-S)%, and that beyond $1{\cdot}25/\sigma$ standard deviations should be σ^2(100-S)/100. The ratio of the two 'single-tail' probabilities should lead to the same value of σ as was used to determine (100-S)%.

May I now, dear Reader, thank you for the patience and, I hope, understanding with which you have followed our progress, from the warnings in Chapter 1 against the dreadful pitfalls of mathematics, to the use in Chapter 2 of this same mathematics to calculate the General Solution, if not to six, at least to four significant figures! You will, I expect, have noted *en route* the probably quite unnecessary references to the limitations of rigorous arguments and natural laws based only on approximations to measurements, and also an invitation, almost on the next page, to accept two of these same laws without reference even to a single measurement in support of them. May I include, in my apology for these two liberties, that for a third, and ask you, in Chapter 3, to consider a little further the perhaps unexpected statistical characteristics of this Neogaussian Distribution with, as yet, but the slightest of references to any 'real' problem. My only excuse is to assure you that this Chapter 3 will be the last of Uncertainty *per se*. Should you decide to persevere with it, I can promise that it will be followed by detailed analyses of two fields of perturbations which are among the most important in Nature, and then by two other detailed analyses of distributions which are of fundamental importance in Communications. In all four of these cases, there will be so much detail and opportunity for direct comparison between theoretical deductions and actual measurements as should satisfy the most exacting of critics. Should you decide to 'skip' this next Chapter and pass on directly to the corroborating evidence, there should be no result other than that some of the consequences of the assumption of Neogaussian Statistics may be the more unexpected, and their 'acceptance' the more difficult therefore.

3

STATISTICAL INTERPRETATIONS

> The chess-board is the world; the pieces are the phenomena of the universe; the rules of the game are what we call the laws of Nature. The player on the other side is hidden from us. We know that his play is always fair, just, and patient. But also we know, to our cost, that he never overlooks a mistake, or makes the smallest allowance for ignorance.

This is, then, perhaps the most unexpected feature of the Neogaussian Distribution, that its parameters are not single-valued as in rigorous mathematical analyses, but sub-distributions of perturbations which conform to particular statistical characteristics. Single-valued parameters occur in the normalised solution, of course, but all of them have their basic uncertainty and serve only to define, to a second order of approximation, the real 'parameters' which are the statistical characteristics of the two causal sets, one (rectangular) toward which the small equiprobable sub-distributions of perturbations tend, and the other (Gaussian) toward which the fluctuations responsible for the coherence in the sub-distributions build up.

The nearly Gaussian statistical characteristic of the larger fluctuations is a consequence of the 'tendency to a mean' of the relatively few coherent fluctuations in the equiprobable sub-distributions, and the effective 'averaging out' of the more numerous perturbations in these sub-distributions. All the Neogaussian normalised probability densities[1] of the sub-distributions in which

[1] Probability density is defined as the relative probability of variate 'values' within a small constant range of deviation in the neighbourhood of any particular deviate.

similar contributions from both causal sets are equally likely are, to some extent (1·095%), indeterminate at all deviates from 0 to 1·0, and the nearly Gaussian statistical law is valid only to this extent and over this finite range. Perturbations always invalidate rigorous laws and limit the range over which they are usefully approximate. By definition the maximum coherent deviate of 1·0 is the point at which the probability density falls to its minimum value at which there is only one equiprobable sub-distribution, in which the frequencies of occurrence are the same for both fluctuations (from the coherent causal set) and perturbations (from the incoherent causal set), so that in the one sub-distribution at this deviate, the effects of the two causal sets become indistinguishable.

There are, of course, many more perturbations than fluctuations in this end sub-distribution (98·905% as opposed to 1·095%), but equal small disturbances in the neighbourhood of any deviate within the total equiprobable range (±0·25) of the sub-distribution are, to within the indeterminacy (1·095%), all equally probable, regardless of causal set.

The fundamental assumptions include two statistical characteristics, and analysis in these circumstances can never be pursued more precisely than the sub-distribution of one probability 'quantum' in which it is impossible to separate their effects. One apple might be distinguished from another in a distribution of apples by position, size, surface reflecting properties, etc. But attempts to define the precise 'position' of an apple, for example (even of its centre of gravity), must depend to some extent on the sub-distribution within each apple, and it can never be specified more precisely than a 'probability quantum' distance which is the basic uncertainty as to which part of the apple defines its position. In everyday life we would almost subconsciously discriminate between the space domain in which the apple is situated and that within each apple, and it is important to do the same in any quantitative analysis also. There must obviously be some uncertainty with which the position of any real object, as opposed to an artificially close approximation to a precise geometrical shape, can be specified. The parameters and laws used to define the position of a real object in a group should surely never be permitted to have the exact values associated either with idealised mathematical concepts or with infinitely large distributions and precise probabilities.

The Gaussian distribution is symmetrical, its probability densities being the same for both positive and negative deviations from its mean. Neogaussian distributions can be derived from either half of the Gaussian distribution separately by adding to it the effects of perturbations. The same normalised solution obviously applies in both cases, but it is unlikely that perturbations in the two cases would be of the same relative importance; the normalisation constant, the point at which small deviate changes are lost in perturbations, is, in general, different.

The whole normalised Neogaussian distribution is defined by three parameters, its mean (0·25), its indeterminacy (1·095%), and the standard deviation (0·4363) of its equiprobable sub-distributions, each of one Neogaussian quantum of probability. It extends (see Figure 1) from the equiprobable sub-distributions having a range 0 to 0·5 centred on zero deviation from the mean of the whole distribution (0·25), to the end sub-distribution centred on a deviate of 1·0 and a variate of 1·25, and extending equiprobably from a variate of 1·0 up to one of 1·5. Each of the 98·905% equiprobable sub-distributions has a probability P(R) equal to the product of the indeterminacy (0·01095) and the variance ($0{\cdot}4363^2$), i.e. to 0·002084. The probability of each sub-distribution is the average indeterminacy over the whole range of coherent deviation ((100-S)%) in the normalised distribution, and is also therefore the average perturbation level over the whole distribution. Those changes which add coherently in successive sub-distributions are obviously unidentifiable in a single such sub-distribution. The ratio of the means of the extreme sub-distributions in the coherent part of the distribution (from a deviate of 0 to one of 1·0) is 1·25/0·25 or 5 to 1, and this fact is often a valuable indication of Neogaussian statistics. If coherent deviates alone are considered, the probability beyond the maximum coherent deviate of 1·0 is 1·095%, and these relatively few 'perturbation-like' fluctuations beyond the maximum deviate may be associated with equally frequent random perturbations. The two sets of small changes will combine randomly on the average, so that all the nearly equiprobable changes from both causal sets beyond the maximum coherent deviation of the complete distribution have an average amplitude range and total frequency of occurrence of $\sqrt{2} \times 1{\cdot}095\%$, or approximately 1·55%. All the coherent changes up to the maximum deviation at which coherence is still discernible (normalised to 1·0) have a total frequency of occurrence, therefore, not of 100% but of only about 98·45% of the whole distribution, which has itself an overall uncertainty of mean of one probability quantum (0·002084).

Statistical characteristics define the frequencies of occurrence of changes, but not the order in which changes occur. It is quite possible for a distribution or sub-distribution of a particular statistical 'shape' also to exhibit ordered cycles of change, or 'waves'. This can result either from some natural periodicity in the time-scale, or from some coherent characteristics in the geometry of the space they occupy. The latter is frequently the case in particle physics and in astronomy, in both of which 'bodies' are often nearly, although not quite, spherical, and often have *mostly* coherent orbital motions and spins. The presence of cycles of changes or waves in Neogaussian statistics is simply an indication of additional coherence, which may affect the way in which the whole field of changes builds up to a total distribution with a mean which is constant apart from its basic uncertainty. It may also affect the total number of identifiable sub-distributions in the whole distribution up to this point.

Disturbances may follow one another in time at a particular place, and/or occur simultaneously but distributed in space. Since all changes take time to occur, both these 'time' and 'space' distributions may often be taken into account together. When this is done and there are no waves or ordered cycles observable, the total number of 'minimum coherence' sub-distributions (each one quantum of probability) in the whole 'randomly' ordered normalised distribution in time and space will be $1/P(R)^2$, or 230,200. This number of sub-distributions, each with a probability of one Neogaussian Quantum, will build up to a total distribution in time and space coherent analysis of which is limited both microscopically and macroscopically by its Basic Uncertainty. In the distribution in space, there is a distance the coherent change over which corresponds to the fluctuation in the equiprobable end sub-distribution of one Probability Quantum (P(R)), at deviation 1·0. Similarly there will be a time-duration of the coherent fluctuation in this sub-distribution in the distribution of changes with time at one place. The ratio of these two is obviously the 'Coherent Mixing Velocity' with which coherent changes spread through the doubly indeterminate distribution. This concept is of considerable value in turbulence and diffusion problems. Since the equiprobable sub-distribution is only 1·095% coherent, there must also be, within each sub-distribution, an 'Incoherent Mixing Velocity', with which perturbations or small incoherent changes must diffuse to maintain the nearly constant mean implicit in 'equiprobability'. This is clearly 100/1·095, or

91·3 times as great as the coherent mixing velocity. Both these mixing velocities change with important change in physical conditions.

Only by considering changes averaged over a time-interval or over a range of deviation larger than those of the sub-distribution of which the probability is one Neogaussian Quantum can the Gaussian coherence of the fluctuations among them be detected. Such 'part distributions' have frequencies of occurrence which become less equiprobable and more and more nearly Gaussian in form as their size increases and their perturbations cancel out. If the distribution in space at any particular time is considered by itself, part-distributions, each consisting of $100/1{\cdot}095 = 91{\cdot}3$ sub-distributions of one probability quantum, will extend over a coherent deviate range from the mean to a maximum of $1/\sqrt{91{\cdot}3} = 0{\cdot}1046$ in the whole normalised distribution. These part-distributions are 98·905% coherent and only 1·095% incoherent. If both the time and space distributions are considered together, the distribution of coherent changes averaged over $\sqrt{91{\cdot}3} = 9{\cdot}56$ times both the time and magnitude range of the 'probability quantum sub-distribution', will be 98·905% coherent and 1·095% incoherent up to a maximum normalised deviation in both time and space distributions of $1/\sqrt[4]{91{\cdot}3} = 0{\cdot}324$.

Sub-distributions of less than a quantum of probability are effectively rectangularly distributed, and their probability density in the whole distribution rises proportionally with the reduction in their probability when only the space or the time-distribution is considered alone; it rises proportionally with the square of the reduction in sub-distribution probability if both time and space distributions are considered together. In either case, their probability densities in the whole distribution remain approximately 'Gaussian' over the whole coherent range of deviation. The statistical characteristics which result when the frequencies of occurrence of such small sub-distributions during very long times are plotted for constant coherent deviation ranges are 'log-normal'; that is to say, they are nearly Gaussian when the very small coherent changes are expressed in logarithmic units.

Since the density of the 'probability quantum' sub-distributions falls from its maximum value at the mean of the whole distribution to its minimum at a deviate of 1·0, where there is but a single such sub-distribution, the coherence in any given small range of deviation near the mean is most easily observable, and it falls to vanishing point (complete incoherence or randomness) at the deviation of 1·0. Over this range of deviation, therefore, the distribution may be said to 'Degrade to Random'. 'Diffraction' is sometimes defined as the disturbance and weakening produced by an obstacle to a field of waves passing close to it. Waves are, of course, systematic cycles of change. In analyses of such conditions, the obstacle has generally an assumed coherent form, and it introduces attenuation (obstacle loss) to any parts of the waves large enough to be coherent; such attenuation, of course, gradually increases with distance round the obstacle. Considered in sufficiently small parts or sub-distributions of the wave-front, however, all real waves are incoherent. Sometimes this incoherence is mostly produced at the source; sometimes it is mostly the result of perturbations or turbulence in the transmitting medium along the path. This latter is the case in microwave radio transmission through the atmosphere. Sometimes, as in long-wave radio propagation, it may be the effect on the whole field of waves re-radiated from the body of the obstacle (the 'ground' wave). In any case, diffraction ends at the 'Change-over' point beyond which there is a 'Random' or 'Scatter' field which is more the result of the perturbations in small parts of the wave-front than of the coherence in much larger parts of them; the obstacle loss rate from then onwards is much less to such a random field than to a coherent field. In idealised Neogaussian conditions this change-over occurs at a probability level, on a normalised scale, of 0·002084; in the units commonly favoured by radio communication engineers, diffraction at sufficiently short wavelengths ends at an average 'loss' of 20 log 0·002084=53·6 dB beyond the 'horizon' at which half the coherent field is obstructed by the obstacle, and the loss is therefore 6 dB. This is considered in more detail in Chapter 7.

When the large-scale fluctuations occur not only with approximately Gaussian frequencies of occurrence but in ordered cycles or waves, they may be divided up and analysed as smaller, approximately rectangular, sub-distributions of perturbations. Alternatively, all equiprobable sub-distributions of perturbations may be regarded as including some small changes which exhibit coherence from sub-distribution to sub-distribution, and thus add up to form waves, these coherent small changes being in no discernible way different from the others when only one sub-distribution is considered. 'Wave' and 'corpuscular' theories conflict only when perturbations are forgotten, and the rigorous assumptions associated either with waves or corpuscles are accepted as valid regardless of the magnitude of the changes involved.

It will no doubt have been observed that many of the characteristics of the Neogaussian distribution are of the same general type as relatively recent innovations in physics. The basic uncertainty, the quantum, the simultaneous validity of wave and corpuscular theories, the probability concept of stable orbits in wave mechanics, are all today widely accepted fundamental physical concepts. What is here perhaps new is the idea that these are all inherent in the particular statistics of the type to which the natural world of which we are part is subject. It has been suggested that coherent laws form the basis of nearly all analyses because of an overriding need for simplicity—because the limitations in our mental capacities are such that, without over-simplification, natural changes in the world around would be far too complex for us to comprehend. This is probably fair comment as far as very small-scale changes in very short time-intervals are concerned. When, however, fields of much larger changes in longer time-intervals are considered, we tend naturally to concentrate on the relatively more easily observed small parts of them, and therefore on their statistical characteristics below the change-over point, where they are very often nearly random. Such apparently predominately incoherent fields of changes are often referred to as 'Turbulence'.

In spite of the apparently chaotic conditions within them, these natural fields of turbulence may still contain a very remarkable degree of large-scale coherence. The earth, although it has a surface which may generally seem to us to be very rough, is as a whole very nearly a sphere, spinning on its axis almost uniformly as it travels at almost a constant speed along a nearly circular orbit round an approximately spherical sun, the centre of a solar system at which is to be found nearly the whole mass of the system. A more nearly perfect example of large-scale and long-term thermal and gravitational coherence and stability would be hard to imagine. Man has so far been confined to an exceedingly thin surface layer of the earth, and even should he become able to venture a little outside it, he will inevitably have to carry 'surface layer' conditions with him to survive.

It is because of this large-scale, almost astronomical, coherence that the over-simplifications of the type commonly employed in rigorous mathematics lead so often to such valuable approximations to the facts, particularly for natural changes averaged over relatively large fractions of the earth's surface layer and large parts of its annual cycle. Natural changes averaged over very small parts within the surface layer and corresponding short time-intervals are, however, so 'random' that rigorous analysis of them is practically impossible. In such circumstances, their large-scale coherence can best be detected among the small-scale changes by observing the patterns (statistical characteristics) of the similar sub-distributions of small-scale changes in time and space. Man has been able to continue his evolutionary process so far and for so long because his world is, as a whole, remarkably coherent and stable, and because he has been endowed with instincts, sensibilities, memory and hereditary tendencies which make him peculiarly proficient at 'pattern recognition', at detecting and adapting himself to the large-scale and long-term coherences which are all that he could possibly comprehend amid the turbulent small-scale complexities of his immediate surroundings anyway. In evolution, the race is won by the most adaptable rather than by the strongest or the most fleet.

PART II

UNCERTAINTY IN NATURE

The world's a scene of changes, and to be
Constant, in Nature were inconstancy.

4

IN THE LOWER ATMOSPHERE

For after the rain when with never a stain
The pavilion of Heaven is bare,
And the winds and sunbeams with their convex gleams
Build up the blue dome of air.

Man has evolved in a world the height of which extends roughly from sea-level to the tops of its mountains, that is to say, in a thin skin about five miles thick on the rough surface of a nearly spherical earth about 8,000 miles in diameter. By far the most important natural changes in this world are those in the lower atmosphere, and any statistical analysis of these must be preceded by deciding:

(a) With what qualities and parameters of the perturbations should the analysis primarily be concerned?

(b) How best can the number of variables be reduced to manageable proportions? There are obviously far too many to include them all, but if the analysis is to cover both the larger-scale fluctuations and the smaller-scale perturbations, the most important causes of change must be grouped into at least two causal sets.

Pointers to the answers to these questions may be obtained by considering three important characteristics of the changes in the lower atmosphere.

1. *Turbulence*

The air is a mechanical mixture of gases, mostly oxygen and nitrogen, held to the surface of the earth by gravity. Although its constituent gases have appreciably different densities, the composition of the air, apart from its water content which may appear as solid, liquid, or vapour, is remarkably constant at all times and everywhere over the surface of the earth, even up to heights many miles above ground, and over an enormous range of pressure. This consistency of composition, which can be the result only of intense and continuous turbulent mixing, is essential to our very existence on earth. Air is said to be 'fresh' only if natural turbulence can replace the oxygen consumed during breathing; we suffocate within a few hours in a confined space from which this turbulence is excluded.

In such conditions fluctuations are best studied in terms of temperature changes, since it is temperature differences which indicate the main (coherent) flow lines in the turbulence. A pressure increase in one volume, for example, results in an increase in the average number of molecules moving into neighbouring volumes, and a decrease in the average distance of penetration (mean free path) before collision. The effects of pressure changes in turbulence thus tend to cancel out of the transport equations.[1]

2. *Horizontal layering*

As a result of this turbulence, measurements of the changes in the lower atmosphere within a few seconds and across a foot or two (the 'smaller-scale' changes) lead to sets of readings which are so random as to make rigorous analysis of them practically impossible. Averaged over longer periods of time and larger distances (the 'larger-scale' changes) the measurements are, as a rule, much steadier and more coherent, the random perturbations across smaller volumes and during shorter time-intervals tending generally almost to cancel one another out of the changes averaged over longer times and larger distances. These latter 'larger-scale' fluctuations are almost always much smaller and have more nearly constant averages in the horizontal direction, that is to say, along lines parallel to idealised smooth earth, than in any other direction. Lines of constant average temperature or pressure are known as isopleths, and the lower atmosphere as a whole is remarkable for the extent to which its isopleths of temperature tend to be horizontal. The air is nearly always 'horizontally layered'. The fact that this layering tends to be parallel, not to the actual surface of the earth but to the theoretical surface of an idealised smooth earth, may be seen in the cloud formations from

[1] For a fuller and more satisfactory explanation of this point the reader is referred to Appendix 2 of Max Born's *Atomic Physics*.

aeroplanes or from the tops of mountains on almost any calm day. It is clearly an indication that the layering and the perturbations along lines of constant average temperature are due to a 'causal set' which is gravitational in origin.

3. *Diurnal cycle heating*

Temperature distributions in the vertical plane are quite different in 'statistical shape' from those in the horizontal direction. The layers are relatively very thin in the direction of their height, with mean temperatures in the vertical profile which very often differ almost unsystematically one from another. Superimposed on these almost unsystematic distributions of layer temperatures there is, moreover, a still large-scale continuous fall of average temperature with height, continuing up to thousands of feet above sea-level and indicative of heating from below and a thermal flow upwards from the earth. The way this heat flow and the consequent temperature changes vary with time, particularly above dry land, indicates that it must be the result of the rotation of the earth in the sun's field of thermal radiation, that is to say, it comes from the diurnal solar heat-cycle.

There are thus four scales of disturbances to be considered. Firstly, measured across sufficiently small height intercepts and time-intervals, each layer temperature in the vertical plane has a 'single' value which is the average of a sub-distribution of equiprobable smaller-scale perturbations; in this sense only is there a single temperature at any particular layer height and during the corresponding short time-interval associated with the measurement. In practice these random perturbations within small parts of a layer height and the corresponding time-intervals are frequently ignored, probably because they are too complex to be analysed. Secondly, over larger height intercepts, a nearly Gaussian distribution of temperature gradients, at any rate over dry land, is generally to be observed. Thirdly, averaged over still larger height intercepts and longer time-intervals, there is the relatively consistent fall of average temperatures (indicative of the heating from below) up to a height of 30,000 to 50,000 ft. above sea-level, the point at which this temperature lapse ceases being known as the Tropopause. The whole of the lower atmosphere up to the Tropopause, to which nearly all our weather changes are confined, is the Troposphere, and since the long-term average temperature of the Troposphere as a whole remains substantially constant from year to year, the heat input to it from the diurnal solar cycle must, on the average over the year, just balance its heat losses by radiation. From one year to another the Troposphere, considered *as a whole* and averaged over a very long time-interval, is in thermal equilibrium, with a 'constant' average 'temperature', by which is here implied a small equiprobable range of variations (the fourth scale) about an almost mathematically constant average.

Statistical analysis of variations in the lower atmosphere is thus best expressed in terms of its temperatures, which determine the main lines of flow in the turbulence, and exhibit both the horizontal layering and the diurnal cycle-changes which are two of its more conspicuous characteristics. The two causal sets on which Neogaussian analysis of the temperature distributions in the troposphere can be based are therefore as follows:

(a) One which alone would result in a large-scale Gaussian distribution of temperature differences between actual layer temperatures and the longer-term mean temperature at that height in the vertical profile.

(b) One which alone would produce the random (equiprobable) and isotropic sub-distributions of perturbations across smaller height intercepts and during shorter time-intervals associated with the 'temperature' of the particular horizontal layering. Since layers are horizontal and lie along lines of constant geopotential, these perturbations are almost certainly gravitational in origin. We shall consider this point in greater detail later.

As explained in Part I, the analysis of the whole distribution from both these causal sets will be considered in equiprobable sub-distributions, or Zones, within which any particular small change is as likely to come from the diurnal cycle as from the gravitational disturbances which produce the random perturbations within short times and small height intercepts. Changes on a smaller scale (either in time or height) than this Zone will come predominantly from the gravitational causal set (b) above. They are the perturbations within particular temperatures distributed equiprobably in any sub-distribution of which the probability in the whole distribution is less than that of the Zone. At the Zone, there is a relatively rapid 'Change-over' to the larger-scale law of the fluctuations (temperature differences or gradients), to a law which is approximately Gaussian in form. Fluctuations in part distributions of greater probability than the Zone appear to come predominantly from the diurnal cycle causal set responsible for the larger-scale temperature differences in the vertical profile. The small-scale perturbations are not usually observed when temperatures are measured in the lower atmosphere, either because the measuring instrument is too slow, or because it is aspirated so that the perturbations are averaged out. Since the same statistical shape applies in both distributions, it will be convenient to consider simultaneously both the temperature changes with time at any one place, and those at various heights at any one time. Although the analysis will include both temperature fluctuations and perturbations within any one temperature, the statistical solution is in terms of the parameters defining the nearly Gaussian larger-scale distributions of temperature differences, although, of course, these parameters will be indeterminate as the result of the perturbations within any one temperature.

Two points of explanation are perhaps worthy of mention at this juncture. Firstly the basic air temperature, that is to say, its temperature without the increases which are the result of the diurnal cycle heating, is determined not by the temperature of the earth below but by the perturbations in the gravitational force which holds the air to the surface of the earth. When, in mountainous areas, the ground rises high above sea-level, its surface becomes much cooler, taking on the temperature of the air at these heights. At high altitudes, the ground does not heat the air to its sea-level temperature—on the contrary, it is the air which cools the surface of the ground to its high altitude temperature. Just as a man's body surface temperature is determined by his clothes, so is the earth's surface temperature determined by its atmosphere. In this analysis we shall assume that the gravitational perturbations and basic air temperature are substantially the same throughout the whole of the relatively very thin skin of air (six to ten miles thick on the surface of an 8,000 mile diameter globe) which constitutes the troposphere.

Secondly, the temperature of a gas volume does not define the velocity of all its molecules. It is merely a measure of the *average* thermal energy condition of the molecules in the volume during the period of time under consideration. Here we shall also be considering the statistics of the temperature perturbations in different parts of the whole volume and time-interval about its average over a longer time and larger height intercept. Each temperature of a volume of air will be considered to be a subdistribution of equiprobable perturbations in smaller volumes, these perturbations being small-scale deviations from the single value usually regarded as the 'temperature' of the whole volume. The statistics of these perturbations include the fluctuations associated with the solar heating from below, and it is these latter which add coherently in the anisotropic turbulence.

The most important of the over-simplifications involved in the assumption that the temperature distribution may be ascribed to only two causal sets are as follows:

(a) It will be assumed that the diurnal cycle has the same average intensity throughout the year over the whole surface of the globe. The 'latitude' effects, as the result of which the diurnal cycle is much less intense in polar regions than at the equator, will, in the first instant, be ignored. It is a relatively easy matter to correct for this over-simplification. For if φ is the latitude of the particular area of the earth's surface under consideration, the solar heating per unit area is approximately proportional to $\cos^2\varphi$, and the heat per unit area per unit time-interval will be proportional to $\cos^3\varphi$. $\text{Cos}^3\varphi$ has its average value ($\frac{1}{2}$) at latitude $37\frac{1}{2}°$, so that the analysis is directed in the first place to conditions at latitude $37\frac{1}{2}°$; corrections for other latitudes will be made later (see Table 1).

(b) The annual seasonal effect which results at any one latitude from the obliquity of the earth's axis to the plane of the ecliptic will also at first be ignored. The solution which is then obtained applies at the equinoxes, and the seasonal effects can be added later by assuming that the latitude varies over the range $\pm 23\frac{1}{2}°$ during the rest of the year.

(c) The air is assumed to be a perfect gas and the effects of change of state of water are ignored. The latent heat of water will, of course, distort the shape of the larger-scale distributions from the theoretical form assumed for them (approximately Gaussian) without adding or subtracting appreciably from the total heat applied to, and the overall heat balance of, the troposphere as a whole. An estimate of this distortion can be made later if desired. It is, however, assumed that all of the thermal energy injected into the troposphere is confined to it and balanced by radiation from it. Practically none escapes either as latent heat or in any other form above the tropopause. Similarly, it is assumed that, on the average over the year, the albedo of the earth and the heat transmitted upwards into the air is constant. All other sources of perturbations or fluctuations are ignored. This includes such small effects as the lunar tide in the atmosphere, coriolis forces, the slowing up of the earth, the additional heat from below through thin spots of the earth's crust, and many others far too numerous even to mention.

All of these assumptions will be regarded as specifying a Standard, Average Turbulent, Radio Atmosphere (SATRA), to which we shall frequently be referring later.

The physical picture consistent with these assumptions is that temperature rises in the ground, coming from the diurnal solar cycle, are, as a result of a mechanism not here explained, transferred to the lowest stratum of air in contact with the ground. This air is consequently set into its additional large-scale coherent turbulent movements (it has already the smaller scale random motion due to its 'gravitational perturbation' temperature), and cycles of temperature changes are carried upwards, losing their coherence (degrading to random) as they ascend to heat the upper layers of the troposphere. As each daily pulse of solar heating occurs, a cycle of temperature changes travels slowly upwards from the ground at the relatively low vertical mixing velocity with which the appropriate larger-scale time and space distributions in the vertical profile build up. As the larger-scale temperature differences diffuse and are degraded into the perturbations within particular temperatures, they heat the air to a temperature above that to which the radiation losses have reduced it during the preceding night. Since the total radiation losses are approximately proportional to the thickness (height) of the air stratum considered, the mean temperature of the air falls steadily with height.

When temperature distributions in the lower atmosphere in dry air conditions (in which none of the solar heat is latent) are analysed in zones, which are the limiting sub-distributions, from the probability point of view, at which coherent temperature difference is detectable, there results the approximately Gaussian distribution of temperatures which has already been described in Part I as a Neogaussian distribution. This hypothetical statistical distribution defines (never quite precisely, of course) the frequency of occurrence of temperature deviations from the mean without in any way involving the order in which these changes occur. Obviously temperature gradients will form cycles which correlate with the diurnal cycle changes much more markedly at the lower altitudes over land (in dry air) than at high altitudes over the sea; this fall in 'coherence' with height can be clearly seen in Figure 2. It is statistically an example of 'degradation to random'.

The parameters of the whole distribution of 'zone' temperatures, including the perturbations within each temperature, are, of course, indeterminate as the result of the perturbations within the sub-distribution (zone) defining each temperature, and the whole distribution of temperatures up to the tropopause, and lasting one year, has a mean temperature which can be constant only so far as the minimum value of this indeterminacy, the Basic Uncertainty of the whole distribution, will allow. This is, of course, inherent in the way we have defined temperature. The general solution of Part I is directly applicable to this hypothetical picture, and once the normalised constants of the time and space distributions have been identified, it should be possible to check the appropriateness of the basic assumptions by the extent of the resulting agreements between theoretical deductions and actual measurements. In this connection we shall here accept the usual statistical standards—that is to say, agreements to within 5% will be regarded as 'significant' and those to within 1% as 'highly significant'. The 'order of magnitude' criterion which seems in recent years to have crept into so many otherwise excellent meteorological analyses will not here be regarded as in any way significant.

Considered in 'zone' height intercepts and time-intervals within which there is an equiprobable sub-distribution of perturbations and a single temperature, therefore, the effect of any further solar heating from below will obviously be to expand the air so that there will then be further random disturbances with a mean coherent flow-path on the average in slightly downward directions, i.e. in the same direction as the bending of the surface of idealised smooth earth, from the heated zones into neighbouring volumes of the layer otherwise in thermal equilibrium with them. Without any such additional heating there will be no such flow and the air would all be at the same temperature. Similar coherent temperature increases in succession in neighbouring volumes of air would build up over longer distances and times to an approximately circular curvature of path flow—this is the only way circular flow-path curvatures can be approached in

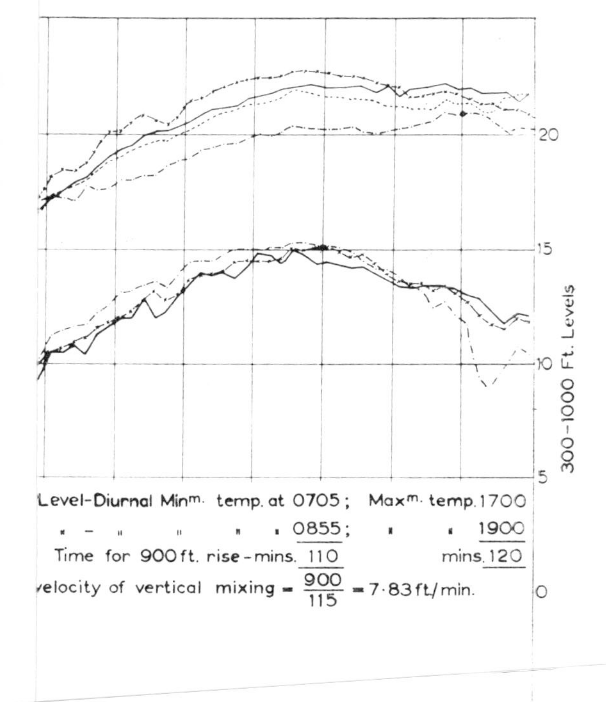

[To face p. 58

viations
a large
circular
rvature'
smooth
here of
f curva-
Straight
rded as
rvature,
d earth

le were
he same
turbu-
eviation
changes
onds to
ter than
heated
layering
easingly
r mean,
ould be
quently
rbations
the case
up over
g there-
t which
at which
tistically
perature
l a unit,
t in the
In the
s clearly
hen the
normal-
ained in
ir hypo-
de 37½°,

and an earth of which the axis is normal to the plane of the ecliptic, i.e. SATRA conditions.

This normalised solution is valid only if the whole distribution is considered in zones each of which is an equiprobable sub-distribution in which small changes are as likely to be isotropic perturbations within volumes at a single temperature as sufficiently small parts of the diurnal cycles of temperature changes rising through those volumes. Air volumes all at the same 'temperature' are, of course, to be found along the isothermals (layers) here assumed to be parallel to smooth earth. The solution has been normalised in terms of the upward flow-path curvatures (upward relative to smooth earth, that is) resulting from the diurnal cycle to which the coherent fluctuations in the turbulent sub-distributions along these horizontal layers add up over heights greater than that of the zone (see Figure 2). Sub-distributions of disturbances are isotropic and equiprobable always, and everywhere only over small height intercepts in the vertical profile and during short time-intervals of the diurnal cycle which can now be calculated from the normalised solution, as follows:

(a) The mean curvature of coherent flow in the turbulence, C_a, is 0·25 curth. This means that, on the average, the coherent diurnal cycle fluctuations in the 'constant temperature' layers add up over greater heights to a mean flow in the turbulence with a path curvature which is $\frac{1}{4}$ of that of the surface of idealised smooth earth; relative to smooth earth, the average direction of flow is, therefore, upwards with a curvature of $-\frac{3}{4}$ curth. The temperature perturbations impose similar bending on the path of radio microwaves, a fact which has been appreciated for very many years by radio engineers, who are in the habit of describing the average *relative* path of radio waves through the atmosphere just above earth by assuming that it is straight (zero curvature) above a 'fictitious earth' of which the radius is $\frac{4}{3}$ of that of true earth and of which the surface curvature is therefore $\frac{3}{4}$ of that of true earth . An average fictitious earth's radius factor $K=\frac{4}{3}$ has been internationally standardised for radio propagation purposes for many decades.

(b) The maximum coherent deviation in this distribution of path curvatures in the turbulence, that at which the distribution 'degrades to random' so that the total probability of all coherent curvatures at larger deviations is no greater than that of similar 'temperature' perturbations in the end-zone at this deviate, is 1·0 curth. At and beyond this deviation, all disturbances are most probably incoherent, and more likely to be isotropic temperature perturbations than diurnal cycle fluctuations. This, of course, is linked directly with the assumption of horizontal layering, and it follows also from this that the maximum *absolute* coherent curvature in the turbulence is 1·25 curth (see Figure 1), at a deviate of 1·0 curth from the mean of 0·25 curth.

(c) The standard deviation of zones (σ) is 0·4363 curth. This is consistent with a maximum Gaussian deviation of 1 curth in S=98·905% of the diurnal cycle. Each zone is thus a sub-distribution of 98·905% isotropic and equiprobable gravitational perturbations, together with 1·095% ((100-S)%) similar but coherent solar cycle fluctuations, which add up over longer time-intervals and larger height intercepts to form the approximately Gaussian distribution of temperature changes assumed for the solar cycle. These latter coherent fluctuations may be regarded as introducing indeterminacy into the equiprobability of the sub-distributions of perturbations, just as the perturbations may be considered to be introducing indeterminacy into the Gaussian characteristics of part distributions larger in height or lasting longer in time than the zone. It depends entirely on which of the two statistical 'laws'—equiprobability or Gaussian—is used as the reference. The frequencies of occurrence of all single-temperatures (zones), and of all the corresponding coherent 'curvature' deviations in the whole distribution, from the mean of 0·25 curth to the maximum of 1·25 curth, are Neogaussian, that is to say, they are Gaussian apart from the fact that they are indeterminate as a result of the gravitational perturbations in them, by (100-S)% or 1·095%.

(d) The zone is the sub distribution of disturbances defined by the probability, $P(R)=\sigma^2\times\frac{(100\text{-}S)}{100}=0{\cdot}002084$, of its coherent component. It is thus 0·002084, or about 1/480th of the whole distribution during one solar day, so that it is a sub-distribution of disturbances which lasts almost exactly three minutes, and in which the coherent component extends over a height which is 1/480th of the height of atmosphere mixed by the diurnal cycle changes rising in the turbulence in one day. Within this 'zone time-interval' and 'zone height intercept', temperatures are apparently 'constant', and the coherent heating of the diurnal cycle cannot be distinguished among the gravitational perturbations other than by averaging over greater height intercepts or longer time-intervals. This zone time and height intercept are thus also the 'Basic Uncertainty' in the time and space distributions of coherent changes.

Since the average direction of turbulent mixing is only $\frac{1}{4}$ of that of idealised smooth earth, it will be, relative to the earth, upwards with a curvature $-\frac{3}{4}$ of that of the earth. This, of course, accords with the common experience that air heated from below tends to rise. There will be a thick stratum of air nearest the ground, therefore, through which the diurnal cycle changes rise during one day. As these cycles of changes rise in the turbulence, the mixing heats this lowest stratum of air to a temperature appreciably above that to which it fell during the preceding night. Turbulent flow from this lowest stratum, which contains a complete but progressively degrading diurnal cycle of heating, will in turn heat a stratum in which there will be similar but mostly incoherent perturbations immediately above it, the average turbulent flow in this second stratum having also a mean curvature of $\frac{1}{4}$ curth *relative to the flow in the lowest stratum* and, therefore, $\frac{1}{2}$ curth in space and $-\frac{1}{2}$ curth relative to smooth earth below. After four such strata of heating one above and after the other, the average direction of turbulent flow will have a curvature of one curth and be parallel to smooth earth, therefore, that is to say, there will then and there be a layer along which there is no longer any predominant *upward* flow component in the turbulence. By the time the mixing has reached this height and at the end of the fourth day of heating, the frequency of occurrence of all the coherent diurnal cycle temperature disturbances coming from below will have fallen to that of similar gravitational perturbations within the layer temperature at this height, the whole distribution having 'degraded to random'. The average temperature at this height will in fact have the 'basic' value due to gravitational disturbances alone, the diurnal cycle heating being balanced by the radiation losses from the whole troposphere up to this height. The height to which this turbulent mixing and the diurnal cycle heating rises in four days and at which the lapse rate of temperature ceases, will here be called the 'Tropopause', although meteorologists generally adopt a slightly different definition for this.

The average curvature of all the single temperature horizontal layers in the troposphere up to the tropopause (one curth) should obviously have the basic uncertainty of the whole distribution, so that the whole troposphere must, on the average, extend over a curvature range of 0·002084 curth. Taking the radius of idealised smooth earth as 3,960 miles, this amounts to 0·002084×3,960× 5,280, or 43,570 ft. In the distribution in time, all of the four days of changes in the vertical profile up to the tropopause are only the 1·095% coherent part of the whole temperature distribution, including perturbations, so that the whole distribution will build up to form a troposphere with a long-term average temperature which is constant, to within its basic uncertainty, in the interval of time of which four days is 1·095%, that is to say, in 4×100/1·095=365·3 days. This, to four significant figures, is one year, and confirms the validity of the assumption of long-term average stability (thermal equilibrium between radiation losses and diurnal cycle input) above a basic temperature determined by the solar tide perturbations of the earth as it rotates on its axis and in orbit round the sun. It has been known for some years that the minute slowing up of the earth can be attributed almost wholly to the radiation losses from the atmosphere, and this fact is a pointer to a direct relationship between the basic temperature which provides the main term in the radiation losses, and the gravitational (solar tide) perturbations during the earth's rotations. Neogaussian statistics applied to these perturbations lead to a value for the ratio between the time-length of the complete annual cycle of temperature fluctuations during the year and that of the diurnal heat cycle, for stable orbit and constant average temperature conditions, which agrees with measured values to within 1/60th part of 1%.

Marked correlation is to be expected in dry air between the diurnal cycle and the temperature gradients in the lowest part of the atmosphere nearest the ground, since the temperature changes in this lowest part come directly from those in the surface, and on many occasions (on cloudless days and in dry air in particular) they will be grouped to form substantially a regular solar cycle of temperature changes, therefore. This can be seen in Figure 2. Contributions to the three higher strata of heating will be the result only of much more random changes, and have much less coherence, and correlate very little with the diurnal cycles observable nearer ground-level. Such coherence as is to be observed is presumably the result of heat carried up from the surface as latent heat and there released to produce additional turbulence in which the coherent temperature changes can be detected.

If the diurnal cycle heating rises to the tropopause at an average height of 43,570 ft., in four days, the coherent vertical mixing component in the turbulence must be $43{,}570/4 \times 24 \times 60$ or 7·564 ft./min. This, of course, assumes uniform mechanical properties for the air (a perfect gas atmosphere) without latent heat effects. Since there are 100/1·095 times as many isotropic disturbances in the equiprobable sub-distributions as there are fluctuations contributing to the coherent vertical mixing velocity component, isotropic temperature perturbations must be spreading with a mixing velocity of $100 \times 7{\cdot}564/1{\cdot}095$ or 690·8 ft./min. This is 7·85 m.p.h., and is the average speed with which small temperature perturbations diffuse isotropically in a perfect gas atmosphere in which there are no latent heat effects or laminar flow (winds are largely laminar). Finally, if the coherent temperature changes from the diurnal cycle rise from the ground at 7·564 ft./min., they will ascend in the turbulent mixing 10,890 ft. in 24 hours, so that this is the average height of the lowest diurnal cycle wave in our hypothetical dry troposphere. The zone height is thus 0·002084 of 10,890 ft., or 22·7 ft., and this is the height across which is the basic uncertainty in the distribution of temperature changes in the vertical profile, that is to say, it is the height intercept within which coherent temperature differences cannot be detected among the perturbations within single temperatures. A zone height of 22·7 ft. is, of course, consistent with the zone time of 3 minutes at the vertical mixing velocity of 7·564 ft./min. Perhaps it would be as well now generally to compare these quantitative deductions from the theory with actual measurements.

In the first place the mean of 0·25 curth is a direct consequence of the assumptions of horizontal layering (curvature 1 curth), and that there are only two sets of causes of temperature disturbances, and that the whole distribution is divided into sub-distributions within which it is equally probable that any particular small disturbance comes from either of the two causal sets. We have already referred to the agreement between this mean value of bending and the value accepted for the average bending of the path of radio waves in the lower atmosphere.

The mean height of the tropopause can be deduced from

information given in a letter from the Meteorological Office dated 18 February 1957. Figures taken from this letter are:

Height of the tropopause for various latitudes, Northern Hemisphere (Pressure-Heights, I.C.A.O. scale, in 1000s ft.)

Latitude	January	April	July	October
90	31·9	31·1	32·3	30·4
80	31·5	31·1	32·8	30·4
70	31·2	29·6	34·6	31·4
60	31·8	31·2	35·5	32·9
50	32·7	33·8	38·0	35·1
40	35·3	{36·9, 53·0}	{40·8, 49·2}	{39·0, 50·0}
30	53·7	54·0	51·4	52·0
20	55·8	55·3	52·0	53·0
10	56·4	56·4	52·4	54·0
0	55·8	56·4	52·4	54·0

In the region 30–40° N. there are often two tropopauses. Two figures are quoted where appropriate.

We shall be discussing the reason for two tropopauses at some latitudes a little later, From the above figures, however, it is possible to obtain the average value for any one latitude throughout the year and then to combine the values for latitudes symmetrically disposed about 37½° to give the grand average value for the troposphere as a whole during the year. We can, for example, average the values for latitudes 20° and 30° to give the approximate value at latitude 25°, and combine this with the figure for latitude 50°; this leads to an average value of 44,150 ft. Similarly we can average the heights for latitudes 10° and 20° and combine this with the figure for latitude 60°; this leads to a figure of 43,630 ft. Finally, we can take latitudes 0° and 10° and combine the result with the figure for latitude 70°; this leads to a tropopause height of 43,210 ft. The three values so obtained have a grand average of 43,660 ft. which is the theoretical value to within about a fifth of 1%. All three values are so close to one another as to confirm that the overall average troposphere height is to be found close to the theoretical latitude of 37½° (see Figure 3).

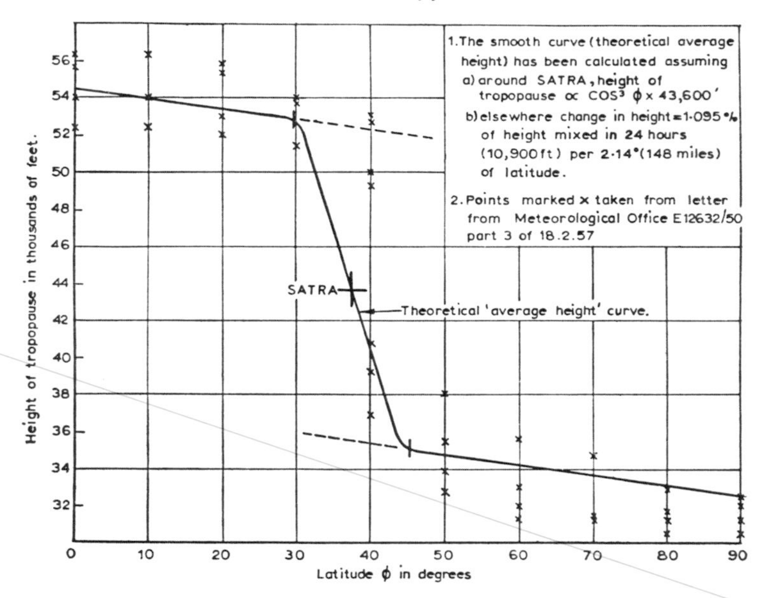

Fig. 3. The height of the tropopause

The relative lengths of the day and the year deduced from the Neogaussian distribution on the assumption that the overall average temperature is constant from year to year is astonishingly close to actual measurements. If the normalised solution is taken to six places of decimals the difference between the theoretical and measured values is, as already pointed out, only about 1/60th part of 1%.

The vertical mixing velocity of 7·564 ft. per minute for the speed with which the temperature gradients of the diurnal cycle rise from the ground in dry air, may be compared with two published figures. In *Geophysical Memoir* No. 89, issued by the Meteorological Office, the time at which the maximum temperature of the diurnal cycle occurred at a height of 107 metres as compared with its time at 15·2 metres above ground, corresponds to an average vertical mixing velocity of 7·53 ft./min. Much more elaborate measurements were made by the staff of the U.S. Navy Electronics Laboratory at San Diego, at Dateland, Arizona, in 1947; the curves shown in Figure 2 were given by them to the author. The average measured vertical mixing velocity in this case is 7·83 ft./min., which is within 1 % of the theoretical figure for this particular latitude (see Table 1, p. 128).

Quite clearly, since the average temperature gradient in each distribution is $\frac{1}{4}$, the average temperature gradient across the whole troposphere should theoretically be $\frac{1}{4}$. J. G. Moore, in the *Meteorological Magazine*, Vol. 85, for 1955–6, has given the average temperature at the tropopause at latitude $37\frac{1}{2}°$ as about 212° absolute, and the average surface temperature at this latitude is about 285° absolute. The theoretical average tropopause temperature is $\frac{3}{4}$ of this latter, or 214° absolute, within 1 % of the measured figure.

If the isotropic mixing velocity is 690·8 ft./min. at the latitude of the average diurnal cycle intensity, then the mechanical wave velocity within small volumes isotropically mixed at the mean temperature should be 100/1·095 of this isotropic mixing velocity, that is to say, it should theoretically be $690{\cdot}8 \times 12 \times 2{\cdot}54 \times 100/1{\cdot}095 \times 60$ cm./sec., or $3{\cdot}206 \times 10^4$ cm./sec. Tables of *Physical and Chemical Constants* (Kaye & Laby, 8th Edition) give the average speed of sound from the Krakatoa volcanic eruption of 1883 as $3{\cdot}21 \times 10^4$ cm./sec.

All these agreements are either significant, highly significant, or even much closer than this. As a final check that the causal set responsible for the Neogaussian 'Perturbations' is the solar tide, K. S. Mitra has given a figure for the solar tide fluctuations in the surface pressure at Washington D.C. as about 1·6 mm. of mercury in 760 mm./mercury, that is to say, it is approximately 0·0021 of the surface pressure, and reasonably close to the value of (P(R)), the 'uncertainty' in this pressure to be expected from Neogaussian statistics.

The above list by no means exhausts the experimental evidence supporting the statistics and the assumptions adopted. We shall be referring to many more such agreements when we consider radio propagation through the lower atmosphere. The extent of the agreement, however, both qualitative and quantitative, must surely justify the acceptance of the theory and its basic assumptions as a means of developing useful second-order approximations to the statistical distributions of the temperature disturbances which actually occur. It remains simply to indicate how the theory may be elaborated to explain some of the more obvious complications in the real atmosphere, and to make a few general deductions which, at any rate in the opinion of the author, seem to follow the acceptance of these ideas.

In the first place, recent information, obtained from the Meteorological Office, about actual heights of the tropopauses (there are, in fact, very often two of them) are shown in Figure 3. The full line in this figure shows the theoretical average heights of tropopause for various latitudes calculated on the assumption that the annual average value of 43,570 ft. occurs at latitude $37\frac{1}{2}°$, and that, in the neighbourhood of this latitude, other values vary as $\cos^3\varphi$, φ being the latitude. In addition, since there is, on the average over the year, clearly no diurnal cycle at the poles, the temperature distribution above the poles must be the result of flow from lower latitudes, so that the average tropopause height over the poles should be less than that at latitude $37\frac{1}{2}°$ approximately by the height coherently mixed during one diurnal cycle. The annual average distribution of temperature disturbances over the poles should theoretically be similar to that between the top of the first diurnal cycle of mixing above latitude $37\frac{1}{2}°$ and the tropopause above latitude $37\frac{1}{2}°$. This leads to a tropopause height of 32,670 ft. at the poles, and since the tropopause height above latitude $37\frac{1}{2}°$ is the average for the world as a whole, its height at the Equator should be 54,470 ft. The resulting smooth curve connecting all these values is shown in Figure 3, and agrees with the average values at all latitudes so well as to be always significant, and sometimes highly significant. It seems likely that discrepancies are due to the effects of moisture. They are most marked at or near the Poles, where ice and snow on the ground would be expected markedly to reduce water vapour pressure in the atmosphere above, and reduce the heat absorbed by the surface, and, therefore, for both reasons, reduce the diurnal cycle heat carried up either as temperature changes or in the form of latent heat to heat layers much above ground-level.

The diurnal cycle heating takes the form of a pulse of heat during daylight hours, but produces practically negligible temperature rises at night. It is therefore scarcely surprising that a tropopause, defined as the height at which the temperature lapse rate, averaged over a kilometre of height (the height mixed in about 7 hours), falls to a very low value, may sometimes occur during what was, at the surface, a night period, after $3\frac{1}{2}$ days of vertical mixing from the ground upwards, and sometimes after $4\frac{1}{2}$ days of mixing starting from periods of time when the diurnal temperatures at ground-level are a maximum. Round about latitude $37\frac{1}{2}°$ there are often two tropopauses, therefore, separated by about 11,000 ft. In latitudes nearer the Poles the frequency of occurrence of the upper tropopause, as defined by a minimum temperature lapse rate over a kilometre of height, would be expected to fall with the fall in average intensity of the diurnal cycle. In lower latitudes, the frequency of occurrence of a tropopause within the fourth cycle of heating (the lower tropopause) should fall markedly as the average intensity of the diurnal cycle heating becomes greater in tropical latitudes. Two tropopauses for these reasons are normally to be expected over the range of latitudes within which the diurnal cycle intensity will give rise to a tropopause height varying between the value to be expected at latitude $37\frac{1}{2}°$, after $3\frac{1}{2}$ days of mixing, and that to be expected there after $4\frac{1}{2}$ days of mixing. More detailed analysis of this explanation would be an interesting study for a statistically minded meteorologist (assuming that there are any such!). The general form of the variations in tropopause height through the year and with latitude, are both such as to support the belief that Neogaussian statistics lead to good approximations to the actual distributions of temperature in the troposphere whenever the effects of moisture are not important (see also p. 125 and Table 1).

When the sun shines on a water surface, most of the energy serves to drive off water vapour and therefore enters the atmosphere as latent heat, there being very small temperature rises in the water itself. This continues until the water vapour pressure near the surface approaches that at saturation. Further heating thereafter will produce an upward component in the turbulent mixing, and conditions will then be very similar to those which occur in dry air, apart from the fact that the zone time taken for any disturbances to rise to the coherent value at which it becomes a detectable temperature change will obviously be longer, and the coherent vertical component of the mixing velocity will therefore be smaller. At some height above ground-level, the average temperature will fall to the point at which condensation is likely to occur. When this happens latent heat will be released and reinforce any other coherent temperature components coming directly from the diurnal cycle, so that the vertical mixing velocity may be nearly doubled. The mean curvature in the part distributions in the height intercepts within which condensation is occurring may be expected to be nearly double the value for dry air, that is to say, nearly 0·5 curth. The effects of the moisture is seriously to distort the Gaussian distribution towards which the larger sub-distributions tend without very much affecting the overall parameters in the atmosphere as a whole.

The only other complication to which we need refer here is that of the obliquity effect. The assumption that the rate of change of diurnal cycle intensity is proportional to the cube of the cosine of the latitude neglects the annual cycle, which is the result of the angle between the axis of rotation of the earth and the plane of the ecliptic. This angle is approximately $23\frac{1}{2}°$, so that during the year the average diurnal cycle intensity may be expected to go through a cycle of change equivalent to latitude changes of $\pm 23\frac{1}{2}°$; the annual range of tropopause height at any one latitude is apparently due to this effect.

More general deductions which seem to follow from the acceptance of doubly indeterminate statistics such as the Neogaussian are:

(a) Values on scales of measurements of temperature, volume, and even distance, force, and energy are not, in the last resort, single valued, even though, in rigorous mathematical analyses, they are generally expressed by single-valued parameters. All such 'qualities' are the result of coherence effects in perturbations within very small volumes and time-intervals, and quantities associated with such qualities should be confined to a range of values determined by the average in the perturbations, on the one hand, and the indeterminacies in the averages and other statistical parameters when conditions over very long periods of time and large amplitudes of change are considered.

(b) Laws expressing quantitative changes in such qualities should similarly be considered valid only over finite ranges of values of the parameters they involve.

(c) Natural changes within the earth's surface 'skin' within which life has evolved can often accurately be explained by assuming Neogaussian statistics, that is to say, by assuming equiprobable perturbations for small-scale disturbances and approximately a Gaussian distribution of larger-scale changes, with stable (constant average) conditions over the skin as a whole and during a very long interval of time. A number of interesting characteristics of coherence-seeking life evolved in such conditions seem to follow from this. We shall be considering some of these when dealing with uncertainty in communications later.

(d) As a consequence of Neogaussian statistics, 'natural' statistical distributions have characteristics which are often linear for very small disturbances and logarithmic for the larger-scale changes. There is a change of law passing from one domain to the other similar to that in the Koenig scale often used for acoustic measurements. Laws such as Fechner's and Weber's may be incomplete in that they do not extend beyond the change-over.

(e) The criterion for the change-over from one law to another is simply the relative frequencies of occurrence of the disturbances in one domain or the other. It is thus purely a probability matter. The maximum amount of 'information' about disturbances in accord with one causal set will be obtained when the whole distribution is analysed in terms of sub-distributions within which disturbances are equally likely to be the result of the other causal set. This sub-distribution is the 'Quantum' of information—it has as its minimum probability the basic uncertainty of the whole distribution. Uncertainty is always introduced whenever the desire for accurate analysis necessitates the inclusion of more than one law—in all natural phenomena it is an obvious limitation to the application of rigorous mathematics. It may sometimes be the result of the indeterminacy produced by the actual measuring process. This is often presumed to be the case in modern physics. It may, however, equally be the result whenever there are two or more sources of disturbances simultaneously operating, as is the case in 'natural' turbulence fields. From the statistical point of view, there is just as strong a case for treating the whole height of the troposphere as the 'quantum' of height along the earth's radius (as far as temperature changes at the surface of the earth are concerned) as there is for regarding the quantum as the minimum detectable element of energy in radiant form.

Our analysis has been in terms of sub-distributions of the particular size (defined by their probability) above which coherent temperature differences from the diurnal cycle start to become observable, and below which they become indistinguishable among the gravitational (solar tide) perturbations within single temperatures. This sub-distribution is the quantum in the scale of temperature *differences* in the lower atmosphere; in smaller sub-distributions only single temperatures can be detected. It contains therefore the minimum amount of information about coherent temperature *differences* to be obtained from measurements, and is the sub-distribution on which any full analysis of them must be based. In SATRA conditions, it lasts for three minutes and extends over 22·7 ft. of height, so that any measurements made on the troposphere will blur over information about temperature differences at some time or another if they are averaged over longer periods of time than this, or apply to temperature conditions over a greater height intercept than this.

Since this sub-distribution is defined by its probability P(R)= 0·002084, the number of them in the whole diurnal cycle is also defined (1/0·002084=480), though of course only with the uncertainty of one sub-distribution. There can be no suggestion of indefinitely large distributions therefore. In the case of the distribution in the troposphere, this limiting stability in the whole distribution of temperature changes has been admitted by assuming that the overall average temperature is constant, that is to say, the average temperature is itself a sub-distribution of perturbations having the quantum of probability. This assumption is valid only because, to a second order of approximation, the earth has a stable orbit round the sun during the year, and because, again to a second order of approximation, its speed of rotation on its own axis is also constant. If the effects of perturbations are not to accumulate from one rotation to the next, this stability involves a theoretical relationship between the rotational periods, one which we have seen agrees with the observed facts as far as the distribution of temperature difference is concerned to within 1/60th part of 1%. There must, of course, be a similar relationship applying to the gravitational perturbations of the earth in its rotations round the sun and on its own axis. There are a number of interesting exercises in the application of Neogaussian statistics in this field which we can now consider in more detail.

5

IN GRAVITY

Observe how system into system runs,
What other planets circle other suns.

Most calculations of the orbital movements of one body (the satellite) around another much larger body (its sun) assume that these movements are completely coherent, and, since this involves neglecting all perturbations, such as are at all times the result of a balance between the centrifugal force on the satellite and the gravitational pull of its sun. When it is assumed that there is always such a balance, the coherent orbital movements of the satellite, excluding perturbations, can often be accurately forecast by applying a relatively simple law, the acceptance of which involves, among others, tne following assumptions:

(a) That the mass of each of the two bodies is a constant which may be considered to be concentrated at a fixed point, its centre of gravity, within it. It is often, but not always, assumed that the gravitational effects of all other masses are negligible.

(b) That for the purposes of calculating the gravitational attraction between them, the distance between the two bodies is the length of the straight line between their centres of gravity, and is unaffected by the rotation of either body, or by perturbations within either body.

(c) That for the purpose of calculating the centrifugal force on the satellite, the only component of its velocity that need be considered is that which contributes to its angular rotation as a whole around its sun.

The heavenly bodies are not perfect, homogeneous spheres, and they have finite sizes even though these may be very small compared with the distances between them. Many of them rotate about only an approximately fixed axis very rapidly, while others are in a state of violent turbulence within themselves. For these, and many other reasons, it should readily be appreciated that no one of these 'coherence-seeking' assumptions is precisely true, and that results of any calculations which depend on them can be accurate only if averaged over velocity changes and time-intervals so large that the overall effects of the perturbations within small parts of them may be neglected. It is, of course, well known that predictions which involve coherent assumptions, and apply over astronomical distances and the times taken for the heavenly bodies to move them, are often remarkably accurate. Averaged over such distances and times, the effects of perturbations are generally very small, and in this sense astronomy is perhaps the most 'coherent' of all the sciences. Nevertheless, we live actually on the surface of one satellite, and its short-term movements so noticeably affect the apparent motions of the other heavenly bodies as to compel us often to apply large numbers of empirical corrections to our 'coherent' calculations to bring them into agreement with observation. Our calendar, of course, is based mostly on empirical rules, and, even so, has needed important readjustments within historical times. It is quite usual practice to assume a number of different, theoretically idealised, astronomical time-scales and orbits for varying purposes, and issue tables (ephemerides) which are, at least in part, empirical, to explain and forecast astronomical events as seen from the earth. Even then, the uncertainties in modern astronomical and geophysical time and distance measurements are relatively easy to detect, despite elaborate expedients introducing second- and third-order coherences to reduce systematic effects to a minimum. The usual coherence-tending assumptions applied in Kepler's laws lead us to expect a simple elliptical orbit for any 'sun-single satellite' combination, and schools often teach that our earth is in such an orbit round the sun. In fact, such a description of the earth's orbit is most misleading, since the departures from an elliptical orbit are often larger than its so-called ellipticity.

Here we shall consider three or four ways in which Neogaussian statistics can be applied to the gravitational disturbances which arise during the earth's rotations on its own axis and in orbit round the sun. In doing so, we shall abandon the concept of a 'coherent' orbit, and assumptions inherent in the use of single-valued parameters in Kepler's laws, in favour of the second-order approximation of admitting perturbations and assuming that only their long-term statistical characteristics can be adequately described using single-valued parameters. This latter implies that the rotations of the earth on its axis and in orbit round the sun are, on the average over a very long time, stable, there being no very long-term accumulations of the effects of the short-term perturbations producing important changes in either.

Both causal sets of perturbations here to be considered arise only because of the earth's rotations on its own axis. Since the earth is not a perfect sphere, it can have no precisely constant axis of symmetry about which to rotate, so that ultimately its movements cannot with certainty be analysed into some which are purely rotational perturbations about a fixed axis, and others which are simply fluctuations in the position of this axis during translation in orbit. Its movements include 'random' components which must be regarded indistinguishably as due both to perturbations in the centrifugal forces on it during its rotations on axis and fluctuations in the forces on it as a whole during its orbit round the sun. The fact that there are random movements of the earth at all shows that these two 'causal sets' are not in short-term balance; the earth's assumed 'stability' applies, to a second order of approximation, only to its relatively long-term average movements.

There must be sufficiently small changes of velocity in sufficiently short time-intervals which are as likely to be the result of axial perturbations during rotation as the result of fluctuations of the forces which, over larger amplitudes of change in longer time-intervals, add to determine the coherent parameters of the statistical characteristics of the earth's orbital velocity changes round the sun. In this particular sub-distribution, small orbital velocity changes (fluctuations) are indistinguishable among similar equally frequent isotropic perturbations which accompany the earth's rotation on its axis. Since forces acting on a satellite produce changes in its velocity, we shall consider as one radial velocity any 'equiprobable' sub-distribution of velocity perturbations up to the limiting sub-distribution, defined by its probability in the whole distribution of velocity disturbances, beyond which a coherent orbital *velocity change* of the earth as a whole is statistically observable, and we shall analyse the whole distribution of disturbances in sub-distributions each of which has this limiting probability. This particular sub-distribution is then the basic uncertainty in the Neogaussian distribution of velocity changes of the earth along its orbit round the sun.

Since the directions of the earth's rotations on its axis and in orbit are both anticlockwise, coherent radial velocity changes outward away from the sun will be regarded as positive. It will be assumed that as a result of one complete rotation of the earth on its axis in one solar day there would be an isotropic and rectangular sub-distribution of velocity disturbances over a finite range which will be normalised to 1·0. The equiprobable sub-distributions of perturbations each lasting one solar day, will add, as a result of the relatively few but indistinguishable coherent orbital velocity fluctuations among them, to form daily cycles of coherent radial velocity changes sometimes (positive) away from the sun and sometimes (negative) towards the sun. As a result of the normalisation, the maximum coherent velocity change will also be 1·0. This is so rare (so far removed from the mean velocity change) that small velocity changes in its neighbourhood occur with a frequency which is no more than that of equal velocity perturbations occurring during the earth's rotations on its own axis. Considered in equiprobable sub-distributions, the normalised solution for the whole Neogaussian distribution of velocity disturbance will then be:

(a) The average coherent velocity change is $\frac{1}{4}$ and along a path which has an average curvature change during one day of $\frac{1}{4}$, relatively to a theoretical unperturbed orbit.

(b) There will be a large number of successive groups of four solar days of velocity disturbances along path curvatures alternately away from and towards the mean orbit within the whole distribution in a long period of time within which the parameters are, to the basic uncertainty of the distribution, constant.

(c) Velocity perturbations within each solar day will form equiprobable sub-distributions during $\pm\frac{1}{4}(\frac{1}{2})$ of a solar day, and each such sub-distribution will include 98·905% rotational perturbations and 1·095% radial velocity fluctuations which add up during longer periods of time to produce the coherent velocity changes.

The calculations are obviously similar to those already applied to the temperature disturbances in the lower atmosphere. Although there are no measurements known to your author (who is certainly no astronomer!) from which a reasonably accurate value of the normalisation constant for the distribution in space can be derived, some of the parameters of the space distribution can be deduced from the consequences of the additional assumptions that both sets of disturbances (in rotation and along the orbit) are the result only of the earth's rotations on its own axis, and that variations in sun to earth distance are negligible. This is statistically equivalent to assuming that the earth's orbit is a perturbed circle, rather than a perturbed ellipse. These deductions are:

(d) If variations in the gravitational pull of the sun are negligible (the distance to the sun being very, very large compared with perturbations in this distance), the long-term balance between solar gravity and centrifugal force must include the average centrifugal force contribution from the earth's rotations on its own axis. There will be, for this reason, an average increase in velocity (and therefore in radius of the mean orbit) equivalent to the average centrifugal force contribution during $\frac{1}{4}$ of a solar day. Disturbances are from a mean circular orbit of this enhanced radius.

(e) The maximum coherent deviation (1·0) in the whole space distribution, from this mean circular orbit to the mean of the sub-distribution at the maximum coherent deviation, is equal to twice the range ($\pm\frac{1}{4}$) of the equiprobable perturbations within the sub-distributions. It therefore follows that:

(1) The time-length of the whole distribution will be 100/1·095 groups each of 4 consecutive solar days of disturbances, or 365·3 days.

(2) Each equiprobable sub-distribution of velocity disturbance lasts for $\pm\frac{1}{4}$ of a solar day, so that the maximum coherent radial disturbance will be that in 98·45% of $\frac{1}{2}$ of a solar day. The maximum coherent deviation from the mean disturbed circular orbit during the year will on the average occur after 100/1·095=91·325 days ($\frac{1}{4}$ of a year), the standard deviation and maximum deviation of the whole of the annual distribution of 'single' velocities (part distributions lasting half of one solar day) being $1/\sqrt{100/1{\cdot}095}=0{\cdot}10464$ times those of the equiprobable perturbations during each half-day. Any coherent ellipticity can only be accurately identified by

averaging out the perturbations in $100/1{\cdot}095=91{\cdot}3$ sub-distributions. Assuming constant average velocity in orbit and radius of the mean circular orbit (to within the basic uncertainty) during the year, the apparent ellipticity (maximum coherent deviation from the mean circular orbit) will therefore be $0{\cdot}9845\times0{\cdot}10464/2\times365{\cdot}3=0{\cdot}000141$. Short-term departures from the unperturbed orbit up to five times as much as this (0·000705) are to be expected in 98·905% of the time. This amounts to $0{\cdot}000705\times360\times60$ or 15·2 minutes of arc.

(3) In the above analysis, the whole distribution of velocity disturbances from unperturbed orbit, through the mean circular orbit on the average $\frac{1}{4}$ of a day's perturbations further from the sun, to the maximum deviation in one day's perturbations, has been treated as a Neogaussian distribution. If the maximum coherent space deviation from the circular orbit during the year is indistinguishable statistically from the average of the perturbations during the day (the nearest approach to a stable circular orbit statistically conceivable), we can regard the daily sub-distributions of perturbations extending outwards from the enhanced mean circular orbit as adding to form during one year a Neogaussian distribution extending over the perturbation range developed in $\frac{3}{4}$ of a day. In this case, as usual, 1·095% of the disturbances will be coherent, and these, combined with random and isotropic components generated as a result of the earth's rotation on its own axis, will account for $\sqrt{2}\times0{\cdot}01095=0{\cdot}0155$ of the whole sub-distribution (see Figure 1). All of the random perturbations developed during rotation about a stationary axis will therefore account for the remaining 0·9845 of the sub-distribution in $\frac{3}{4}$ of a solar day, so that the whole distribution of such sub-distributions will become stable to within its basic uncertainty in 1/P(R) such sub-distributions or $3/4\times0{\cdot}002084\times0{\cdot}9845=365{\cdot}49$ solar days. This is the time-length of the distribution of perturbations (resulting only from the rotation of the earth on its own axis) beyond the enhanced mean circular orbit the radius of which includes the average coherent value in the sub-distribution, that

generated during one $\frac{1}{4}$ of a solar day. It is the time-length of the whole distribution of perturbations consistent with a mean orbit which is constant to the basic uncertainty of the whole distribution, and it is therefore the orbit of maximum long-term stability. If the effects of all perturbations are averaged out, including the enhancement of the average orbit, there will result a year of apparently 365·24 days, which is precisely the length of the average calendar year. Our additional assumption, that the so-called 'ellipticity' of the annual orbit is equal to the average in the equiprobable perturbations during the day and that there is no other source of ellipticity, is thus apparently correct to 5 significant figures. To a second order of approximation, the earth's mean orbit is circular in the sense that its ellipticity is simply the mean of its axial perturbations from its average orbit.

The Neogaussian distribution may also be applied to two systems of gravitational perturbations at the earth's surface. It was shown in Chapter 4 that the diurnal cycles of heating rise in the troposphere with an average vertical mixing velocity of 7·564 ft./min. This, of course, assumes uniform mechanical properties for the air (a perfect gas atmosphere) without latent heat effects. Since there are 100/1·095 times as many isotropic disturbances in each equiprobable sub-distribution within single temperatures as there are fluctuations contributing to the coherent vertical mixing velocity component, isotropic perturbations within single temperatures must be spreading from one layer to another and along individual layers with a mixing velocity of 7·564 × 100/1·095 or 690·8 ft./min. Both of these velocities were deduced in Chapter 4. This isotropic mixing velocity is the speed with which the temperature disturbances maintaining uniform temperatures along layers are disseminated. This isotropic mixing results from the gravitational perturbations associated with the solar tide and these go through one complete cycle in 24 hours, developing their average coherent effect in a quarter of this, that is to say, during six hours. Isotropic mixing should, therefore, on the average, extend from the earth's surface up to the height reached in six hours at 690·8 ft./min., that is to say, up to a height of almost exactly 47·1 miles. It is now commonly accepted that the dissociation of the atoms in the lower atmosphere into ionised particles and free electrons in the ionosphere is neutralised nearer the earth's surface by the isotropic mixing, so that this height of 47·1 miles is the average height at which the ionosphere and the free electrons in it should first be observed. In fact an average height of $47\frac{1}{2}$ miles to the base of the ionosphere is probably as accurate an estimate as can at present be obtained from published data. The figures are consistent with the Taylor–Pekeris theory of solar tide resonance, since the time for the return journey from the ground to the ionosphere and back would then theoretically be 12 hours, and there should be observed a resonance phenomenon similar to that in a short-circuited quarter wave line with a marked semi-diurnal component in the wave.

It is also possible to use the results of the statistical analysis of gravitational perturbations near the surface of the earth (coming mostly from the solar tide) to estimate the ellipticity of the orbit of the earth round the sun. These gravitational perturbations, in association with the earth's main field, result in sub-distributions of perturbations with a mean average force of attraction which is constant along horizontal lines concentric with the earth having an average radius of curvature below the ionosphere close to $3960+\frac{1}{4}(47)$ or 3,972 miles. These sub-distributions of perturbations over small height intercepts and in small volumes along the horizontal, constant geopotential, lines are random and isotropic, apart from an additional coherent vertical mixing component which, over larger height intercepts, is the source of their coherence, the latter manifesting itself only over height intercepts more than 47·1 miles. If as before it is assumed that the so-called ellipticity of the earth's orbit is simply the average effects of its perturbations (i.e. that the earth's orbit is a perturbed circle rather than an ellipse), then the coherent radial outward components acting on the earth during one quarter of a year should bear the same statistical relationship to the coherent pull of the sun as do the gravitational perturbations at the earth's surface during one quarter of a day to the coherent force of attraction of the earth near its surface. Forces lead to accelerations (velocity changes per unit time), and since these latter will occur both in the distribution in space round the orbit and in time during each rotation on axis, the maximum coherent change in dimensions of the earth's orbit during the year should be proportional to $(47{\cdot}1/3972)^2$. This leads to an ellipticity of 0·0001406, and, compared with a measured figure of 0·0001403, is in agreement with it to within $\frac{1}{4}$ of 1%.

Lastly, if superimposed on each mean value of the earth's gravitational attraction there are the solar tide isotropic perturbations extending on the average up to 47·1 miles, then one could visualise a Neogaussian distribution of 'single values' (sub-distributions of equiprobable perturbations) of the earth's gravitational pull all the way from this sub-distribution nearest to the earth's surface to a point much more distant in space above which coherent change in the gravitational field due to the earth has a value which is statistically indistinguishable among the solar tide perturbations. The whole distribution up to this height should then extend for 1/P(R) such sub-distributions, and the smallest coherent value of change in the earth's gravitational pull should therefore cease to be observable above a height 98·45% of 1/P(R) times 47·1 miles. The apparent height at which there is no discernible coherent change detectable among the isotropic perturbations associated with the solar tide should therefore be $0{\cdot}9845 \times 47{\cdot}1/0{\cdot}002084$ or 22,250 miles, with an uncertainty of 47·1 miles. The satellite subject predominately to these two causal sets of change, the earth's gravitational pull and the solar tide perturbations, is the one which travels synchronously with the earth in a 24-hour orbit. 'Stationary' satellites of this kind may be set at the height calculated as that at which the mean gravitational pull of the earth is theoretically exactly balanced by the centrifugal force of the satellite. Assuming that the mass of the earth is $5{\cdot}975 \times 10^{27}$ gm., that its mean radius is 6,371 kilometres, and that the gravitational constant is $6{\cdot}673 \times 10^{-8}$, the height from ground to the orbit of this 'stationary' satellite can be calculated in this way to be 22,244 miles. The values obtained by assuming a stable orbit with minimum perturbations and the value obtained by neglecting all perturbations thus differ one from the other by less than 10 miles in about 22,000 miles, which is less than the uncertainty involved in the former calculation (47·1 miles). It may not at present be possible, therefore, to decide which is the more accurate set of assumptions. The measured average value likely at the moment to be the most accurate is probably 22,250 miles, and variations from this average so far recorded cover a range of about 43 miles.

The success of these applications of the normalised Neogaussian distribution to the temperature and gravitational perturbations close to the earth's surface leads to two general conclusions, or, as will be suggested later, perhaps to two facets of one general conclusion:

(a) Averaged over very large distances and long times, important causal sets of change operating on our world from without are remarkably stable. This permits us to express the distributions of changes they produce using two statistical characteristics defined by a few single-valued parameters. As an example, the temperature gradients (rises) of the diurnal solar cycle are superimposed on a basic temperature which is due to the gravitational perturbations associated with the solar tide. These latter depend on the speed of rotation of the earth and the distance from earth to sun, and both of these have average values which remain substantially constant for very long periods of time. Even our semi-rigorous methods of analysis would have failed in less stable conditions.

(b) Our mental processes are such that we concentrate almost exclusively on the coherent components in our observations, tending subconsciously and consciously to disregard the random changes which always accompany them. Thus we 'see' waves moving horizontally along the surface of water, even when we know that small parts of the water surface are really moving almost vertically up and down. We draw smooth coherent curves, or even straight lines, through clusters of points on our graphs even when very few of our measurements actually fall on such smooth lines. We almost always make or control our machines in the same way. Look up at the sky on almost any clear day—how easy it is to distinguish between natural clouds and man-made vapour trails simply because the latter follow coherent tracks. We manufacture for ourselves an artificial world of nearly straight lines and circles, of cylinders, shafts, and almost circular wheels, of electrical circuits composed of resistances, inductances, and other idealised elements, of 'pure' materials or simple alloys, of flat or smoothly curved roads—all of which have the one essential characteristic that they approximately conform to some coherent design simple enough for us to understand and use in plans and formulae. In any natural environment in which changes are uncontrolled and too complex for us wholly to comprehend, we find that we can most easily gain some insight into what is happening by concentrating only on the coherent components in the changes, by 'pattern recognition', as it is generally called.

> In analysing fields of natural changes, we should therefore make or group our measurements in such a way that the patterning or coherence in them is apparent. In geophysics and astronomy, this may involve measuring over distances and times much larger than is usual in dealing with changes local to us on earth. Conditions in which changes appear mostly random and incoherent are said to be turbulent, and too often we fail in our attempts to analyse them because our measurements have been made over times or distances selected simply for their convenience, without regard to the effects such selection may have on the 'information' content of the data so obtained.

When we do recognise some correlation between our observations and a rule, code, or language with which we have previously made ourselves familiar, we are said to acquire 'information'. This can occur only if there are coherent components (signals) statistically detectable among the accompanying perturbations (noise). We can recover the maximum amount of information only if we measure and analyse in such a way as to include the minimum coherent change which is statistically observable. This maximum amount of information is extracted, therefore, when we analyse in sub-distributions within which small coherent changes and equal perturbations are just equally probable. The probability of such a sub-distribution is then the natural unit of information, just as is the 'bit' in analyses of binary systems in which perturbations are neglected.

Only by analysing up from this particular sub-distribution, which considered alone is simply noise or perturbation, to the larger and/or longer lasting part-distributions, and through them up to the whole distribution limited by its basic uncertainty, can we hope to extract all of the information available in the experimental data. In the Neogaussian distribution we do this by dividing the whole field of changes (or as much of its variate range as possible) into equiprobable sub-distributions or 'zones' within which any small change is as likely to be coherent fluctuation as a random perturbation. The information content of any particular coherent range of the variate is then simply the relative frequency of occurrence (probability density) of the zones in that range; it is most conveniently expressed in logarithmic units. This information has, of course, a constant percentage 'lack of precision' (indeterminacy) as a result of the perturbations. In logarithmic units, there is a constant 'imprecision' or blurring over the whole information range of coherent deviation. In the end, at the 'tail' of the distribution, the probability of all the theoretically remaining coherent fluctuations falls below that of the perturbations within the zone, so that there can be no further pattern recognition of zones. Beyond the end-zone, the total probability of coherent change falls to the minimum indeterminacy in single zones, i.e. to the Basic Uncertainty, which sets the limits to the extent, information range, and precision of the whole analysis. The relationship between information, indeterminacy, and basic uncertainty is, perhaps, in its simplest form in the Neogaussian distribution, but something of the sort would seem to be inevitable in any 'explanation' of complex natural phenomena satisfying an intelligence able to recognise only relative simple statistical patterns.

Many animals habitually make patterns in their movements or sounds which in some way derive from their past experience or heredity, and thus pass on information about their condition or circumstances to others of their kind with similar experience or heredity. In this way they can transmit one to another a sense of fear, or information on where food is to be found, and a great deal of other currently useful information. It appears, however, that only in man is there the ability to transmit information completely divorced from his environment or physical experience at the time of the transmission. He can record and transmit abstract ideas, choosing when, where, and what to transmit regardless of his immediate environment and stimulation. To do so, of course, he uses a language of coherent sounds, codes of signals, and alphabets of visible patterns enormously more elaborate than any used by other forms of life. In particular, he alone can speak, write and 'telecommunicate'. When he speaks he produces patterns of coherent, meaningful sounds which convey information in a statistically turbulent field of sound changes which acts as a carrier. Let us now see how the Neogaussian distribution can be used to explain events in this, by far the most important of all our means of communication.

PART III

UNCERTAINTY IN COMMUNICATION

We'vc trod the maze of error round,
Long wandering in the winding glade;
And now the torch of truth is found,
It only shows us where we strayed.

6

IN SPEECH

The mind, that ocean where each kind,
does straight its own resemblance find.

It has now surely been confirmed that the statistics of fields of natural disturbances in the world around us can be expressed accurately by assuming that they are the results of two causal sets, one of which leads to very large numbers of small equiprobable perturbations within sub-distributions or 'single values' of the more obvious parameters of temperature, or pressure, or gravitational force, the other causal set being responsible for the larger-scale coherent distributions of the different values of temperature, or pressure, etc. Apart from the fact that the statistical characteristics of the perturbations must be close to what we have assumed for them, very little is as a rule known about the reasons for particular values of them, or about the individual causes contributing to their causal 'set'. Our measurements are in most cases carefully arranged so that these complex disturbances are averaged out of them. The coherent pattern produced by the larger-scale coherent causal set, however, can often be easily identified, as, for example, the solar cycle of temperature gradients which is their main cause; most of our measurements and sensibilities concentrate particularly on identifying such coherent patterns. Even though we may be quite unable to extract all of the information available in the tens or hundreds of thousands of nerves stimulated by incoming perturbations, we can readily accept information about the patterns in them due to the larger-scale changes which result from their coherence. When we do accept and pass on 'information' from one to another by pattern recognition in this way, either directly or via some man-made transmission system, we are said to 'communicate' one with another. Looked at in this light, it seems natural to suppose that the sensibilities through which we receive such communications have evolved in the way most suited to their natural surroundings, so that the same pattern recognition process and its associated Neogaussian statistics should be applicable to the analysis of perturbations in speech and in visual fields as to disturbances in the atmosphere around us. We can now test this hypothesis by applying Neogaussian statistics to speech 'waves'.

George Bernard Shaw once remarked that he could tell what a man had had for dinner merely by listening to him on the wireless. He also credited one of his more important characters with the ability to identify the place of origin of a speaker, to within a mile or so, simply by listening to his speech. While there is, no doubt, an element of poetic licence in such claims, the fact does remain that there is nearly always much more information to be obtained from speech than is conveyed by all of its words taken one by one. As has already been suggested, information is acquired via a pattern recognition process, and this latter is a question of scale. There can be smaller, meaningful, patterns which form part of larger and longer-lasting patterns which, in their own right, add further information. Sometimes the contribution of many unspoken words is conveyed by the context. The question 'How much?', for example, may contain the whole significance of 'How much money does this cost?'. In 'Neogaussian' analyses, one of the two causal sets of changes is assumed to be an incoherent and informationless cause of perturbations, and we do best therefore to take as the coherent causal set the source of the smallest identifiable patterns. In speech, these latter are syllables, rather than words or sentences. We can as a rule attach no particular significance to the separate parts of a syllable, which is therefore the smallest information bearing pattern, but that we can do so to whole syllables is clearly evidenced by the fact that there are many single syllables which are also complete words. Of course, if we base our pattern recognition on syllables, we shall be restricting ourselves to part only of the total available information, rejecting that part contributed by groups of syllables (most words) and groups of words (phrases and sentences). Here, the restricted information which can be derived from syllables alone will, following telephone practice, be called Articulation, a term used also to refer to the actions of the jaws, tongue, lips, and other articulating organs during speech. The minimum detectable amount of articulation (information), the 'zero' of a logarithmic information scale, is that in the sub-distribution of small pressure changes within which those coming from articulation (actions) and contributing to articulation (information) are no more probable than the equal pressure perturbations, also to be found in speech, which come mostly from the edge-vibrations of the vocal cords. This equiprobable sub-distribution, with the indistinguishable coherent component responsible for its information content, will be called the *QUIP*, a name which, it is hoped, will remind its user that it refers to the e*QUIP*robable sub-distribution which contains the *QU*antum of *I*nformation among the *P*ressure perturbations.

In our discussion it may be necessary to refer to the following parts of the upper respiratory tract involved in the act of speaking:

The Larynx is the cavity in the throat containing the *Vocal Cords* and the *Glottis*, which is the space between them. It is continuous with the upper part of the windpipe, so that movement of the vocal cords is associated with pressure changes in the breath stream during exhalation. It includes what is popularly called the 'Adam's Apple'.

The esophagus (or œsophagus) is the tube from the stomach via the larynx to the *Pharynx*, the latter being the cavity which terminates it as the back of the nose and mouth.

The Articulating Organs are the tongue, jaws, palates and other parts of the mouth set into controlled movement during speech. The general arrangement of all these parts is shown in Figure 4.

The physical picture at the listening 'end' is as follows:

Speech sounds enter the ears as a continuous train of small pressure changes, and pass through a tube, the *External Ear Canal*, the cross-section of which is too small adequately to transmit any recognisable (coherent) part of the incoming sound wave-fronts. After traversing the ossicles and oval window in the ear, this train of small pressure changes impinges on one end of the *Basilar Membrane* in the Cochlea, which is connected to the brain by a very large number of nerve fibres (there are normally more than twenty thousand of them). Pressure changes travelling along the basilar membrane stimulate the nerves attached to it, and these in turn develop sensation in the brain. The arrangement effectively converts the succession of small pressure changes entering the ears one after another in time, into a series of space patterns of activated

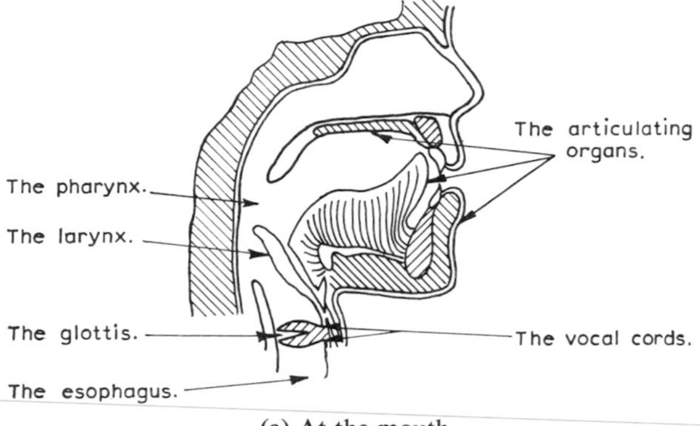

(a) At the mouth

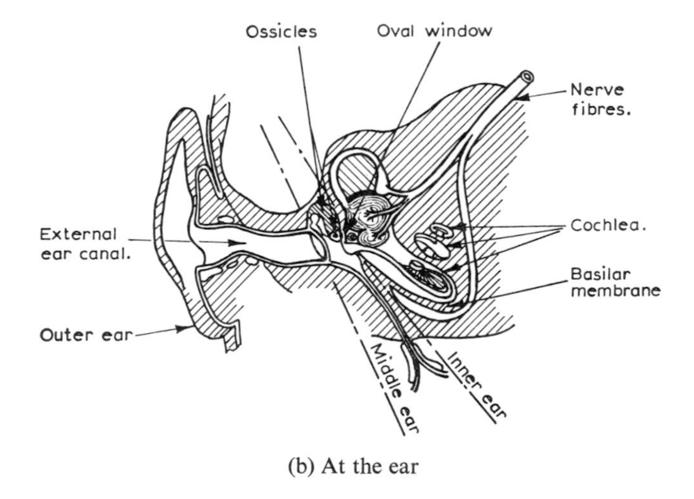

(b) At the ear

Fig. 4. General arrangement of the organs used in speech

nerves along the basilar membrane transmitted to the brain of the listener. 'Recognising' speech consists essentially of identifying these patterns as similar to those which the listener can develop in his own ears by using his tongue, jaws, and other articulating organs to speak.

Speech is thus essentially a 'modulated carrier' type of transmission, the carrier consisting of the perturbations in the breath stream produced mostly at the edges of the vocal cords in the larynx during exhalation, in much the same way as are perturbations produced in an air stream when air is forced past the edges of the opening in an organ pipe. During speech, the vocal cords as a whole may be tensioned in such a way that, with their connecting passages, they have also a relatively low audible fundamental frequency of their own, just as has an organ pipe. In general, this larynx tone varies from person to person and from time to time in the same person during *voiced* articulation, and though it can be carefully controlled as in singing, such tensioning does not as a rule play an important part in conveying articulation information, at any rate in Western European languages. Nevertheless, when there is a recognisable larynx tone in voiced speech, it should be possible to identify it as the almost entirely coherent part-distribution which results when the appropriate number of quips are considered as a single entity, so that the random perturbations in them mostly average out. This larger coherent part-distribution will here be called the *Colar*, a name derived from the words coherent and larynx.

With this physical picture in mind, we can now apply Neogaussian statistics to describe the distribution of small pressure increases in a continuous, hypothetical speech at a point in air close to the mouth under conditions in which the effects of external environment, including those of any microphone or telephone receiver, are negligible. The two causal sets are, of course:

(a) that which *alone* would produce an equiprobable distribution of small pressure changes (random perturbations) without any coherent component among them. We shall consider these perturbations as due to the edge-vibrations of the vocal cords as air is uniformly expelled through the glottis, although there is no precise or exclusive assignation of their source either necessary or intended here.

(b) that which *alone* would produce a Gaussian distribution of generally larger and longer lasting pressure changes (coherent fluctuations) above atmospheric pressure; small parts of these, with the perturbations (a) above, form the equiprobable sub-distributions on which we shall base our analysis. We shall refer to this as articulation, although some of the coherent movements of the articulating organs, notably during sibilants, may also result in random perturbations in the air-stream at their edges.

We shall assume that the average level of speech is practically constant over periods of minutes. This is implicit in the approximation that there are only two causal sets.

In these circumstances, the frequency of occurrence of small coherent (articulation) pressure rises gets less as we depart more and more from the Gaussian mean pressure rise, until, at a maximum coherent deviation, there will be a small pressure change the frequency of occurrence of which is the same as that of equal perturbations coming from the edges of the vocal cords ((a) above). As usual, we can normalise by choosing as one the fluctuation and perturbation ranges so that the maximum coherent deviation p_n is equal to the total pressure range over which the equiprobable perturbations from (a) extend in their whole distribution.

The normalised Neogaussian solution for this particular case is then as follows:

1. The mean pressure rise (at zero coherent deviation) of the whole distribution of quips in our hypothetical speech is $0{\cdot}25\ p_n$.

2. The maximum coherent deviation from this mean pressure rise is p_n, and, beyond it, above a pressure rise of $1{\cdot}25\ p_n$, the total probability of all coherent fluctuations (pressure changes coming from the articulating organs) falls to the indeterminacy due to the perturbations (1·095% in the whole distribution up to this point).

3. The equiprobable sub-distribution, the quip, has a probability of 0·002084, and in it there is a coherent pressure fluctuation, the minimum detectable articulation at the deviate p_n, of $p_q = 0{\cdot}002084\ p_n$.

4. The whole distribution is equiprobable in the neighbourhood of all its coherent deviates over a pressure range of $\pm 0{\cdot}25\ p_n$. Quips occur over the pressure rise range from $0{\cdot}25\ p_n$ to $1{\cdot}25\ p_n$ with Gaussian probability densities apart from the indeterminacy in the whole distribution, which amounts to 1·095%. Each quip includes the minimum coherent fluctuation at the mean $0{\cdot}25\ p_n$, and the maximum coherent fluctuation at the maximum coherent deviation of p_n (total pressure rise $1{\cdot}25\ p_n$) where there is a single quip extending from p_n to $1{\cdot}5\ p_n$.

5. The total probability beyond this end quip, i.e. beyond $1{\cdot}5\ p_n$, is 0·002084. This is therefore the basic uncertainty in the whole distribution.

6. The standard deviation of the whole distribution of $1/0{\cdot}002084 \approx 480$ quips is $0{\cdot}4363\ p_n$.

If we assume that the medium in which the speech travels is uniform, so that its mixing velocities are everywhere and at all times the same, the statistical shape of the distribution in time will be similar to that in space. Sub-distributions lasting longer, and involving more coherent pressure change than the quip, will be increasingly coherent, until in $\sqrt{100/1{\cdot}095}=9{\cdot}556$ times the quip time t_q, during which the coherent pressure rise change will average $9{\cdot}556\ p_q$, there will be a larger sub-distribution or part distribution containing $100/1{\cdot}095=91{\cdot}33$ quips which, considered as an entity, will be 98·905% coherent. This is the colar. The external ear canal is physically too small to accept any recognisable coherent part of this in the distribution in space along the wave-front external to the ear, but of course we can identify it quite readily as a frequency by identifying the larger mostly coherent pressure changes which occur one after the other in the distribution in time.

Many published measurements of speech are either in terms of its frequency characteristics or of its pressures. In this statistical analysis of course we shall refuse to admit the existence of precise values of pressure or frequency. As far as we are concerned there can be no single-valued parameter describing the whole distribution. Pattern recognition is possible only at probability levels above that of the perturbations, and exact, single values can have no probability. We shall, however, refer to pressure changes or pressure gradients, meaning in the latter case not an unrealistic 'gradient at a point' but the average rate of change over a small but finite pressure change, or during a short time-interval.

In order to apply the normalised Neogaussian solution, we shall need to take from our published measurements two normalising values, one for the distribution of pressure changes or gradients one after the other in time (perhaps very near the ear at the entrance to the external ear canal) and the other for their distribution in space, i.e. in the syllable pattern as it extends, for example, at any one time along the basilar membranes in our ears. Values used here for the distribution in time have been taken from the published data which it is believed incorporate the largest number of controlled measurements made in the English (or American) language. These measurements are those made by the Bell Telephone Laboratories in the United States and published in 'The Perception of Speech and its Relations to Telephony', Harvey Fletcher & R. H. Galt, *Journal of the Acoustical Society of America*, March 1950, and 'A New Frequency Scale for Acoustic Measurements', W. Koenig, *Bell Laboratories Record*, Vol. 27, No. 8, August 1949.

The two curves in Figure 4 of Fletcher & Galt's paper have been reproduced in Figure 5 attached. They cross at a frequency close to 1830 c/s, which is therefore the mean frequency from the articulation point of view, there being as much articulation in speech limited to all frequencies below this cross-over frequency as in speech restricted to all frequencies above it. According to our hypothesis, contributions to articulation information are the result of the coherent pressure changes in sub-distributions (quips) in which similar pressure changes from the incoherent causal set (edge vibrations of the vocal cords) and from the articulating organs are equally probable, so that 1830 c/s is the frequency associated with the average quip of the whole distribution. The time-interval during which the coherent pressure change in this

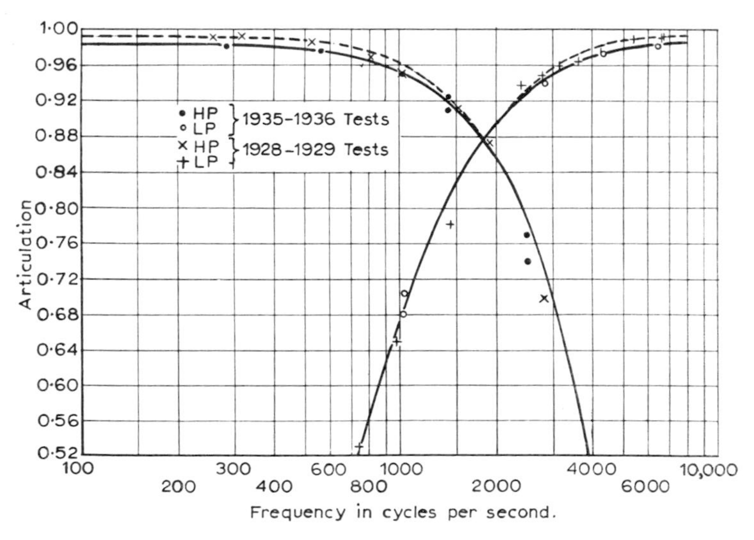

Fig. 5. Speech sound articulation *vs.* ideal filter cut-off frequency

average quip at 1830 c/s builds up is clearly a half-cycle of this frequency, that is to say the quip time t_q is $1{,}000/2\times1830$ or 0·273 m./sec. This is the one empirical figure we need to build up the whole distribution in time.

As a first step we can check that the shape of the curves in Figure 5 is consistent with our assumption of Neogaussian statistics, observing that articulation should be recognisable up to a maximum coherent deviation beyond which the total probability of larger coherent gradients is 1·095% and that of similar incoherent changes, the indeterminacy at this deviation, is also 1·095%. This is the total probability under the tail of the Neogaussian curve beyond the middle of the end sub-distribution (see Figure 1). That part of the total distribution beyond which there is no discernible pattern and coherence is thus, on the average, $\sqrt{2}\times$ 1·095% or 1·55%, and the theoretical maximum articulation of speech as a result of the pattern recognition process is therefore 98·45%. It will be seen that the maxima of the two curves of Figure 5 are approximately 98·5%, and the figure given by Fletcher & Galt in their text for the maximum measurable articulation is also 98·5%, so that theory and measurement are reasonably well in agreement here. If we divide the whole distribution, the total articulation of which is theoretically 98·45%, into two equal-articulation halves, the maximum articulation in each half, as a result of the pattern recognition process, will be less than 100% by $\sqrt{0{\cdot}0155}$, i.e. less than 100% by 12·44%. The two curves of Figure 5 do in fact cross over close to this point (articulation 87·6%) and their general shape is otherwise much as would be expected from an assumption of Neogaussian statistics.

If the quip time t_q is 0·273 m./sec. our hypothetical syllable time should be 0·273/0·002084, or approximately 131 m./sec. The Bell system articulation measurements were based on the use of logatomes or separate syllables, and it would seem reasonable in these circumstances to assume that any continuous speech composed of such sounds might be regarded as a series of time-intervals in which, on the average, the time taken to recover from each syllable in preparation for the next is equal to the time-length of each syllable, and the pressure gradients developed during the gap between syllables, that is to say during the recovery action, are random with respect to those during the actual syllable itself. On this assumption we can add 1/0·002084 or 480 (syllable+gaps) to form a hypothetical '*call*', lasting $2\times131\times480/60\times1{,}000$, or

2·1 min., in which the speech volume is constant to within the uncertainty of one (syllable+gap). This uncertainty is therefore about a quarter of a second.

We can also identify an interval of time, the colar time t_c, in which the coherent pressure changes in quips add up to form a 98·905% coherent part of the syllable. This will occur after $\sqrt{100/1{\cdot}095}$ or 9·56 quips, that is to say, in 2·61 m./sec. The almost entirely coherent pressure rise occurring in 2·61 m./sec. would appear as a fundamental larynx frequency of 1000/2×2·61, or about 190 c/s. Fletcher has reported elsewhere that the larynx tone for men was about 125 c/s and for women 244 c/s, so that our hypothetical Neogaussian speech has an average larynx tone close to that of the adult population.

We can also apply Neogaussian statistics to calculate the contributions of small parts of the frequency band to articulation. Narrow frequency bands, each having a width which is a constant percentage of their mid-band frequency, will include all pressure gradients in speech which fall within a constant small percentage of a value determined by the mid-band frequency, whether these pressure gradients are the result of small pressure changes along the space-pattern which extends at any one time along the sensitive membrane in the ear, or whether they have arisen as a result of pressure changes one after the other in the time-pattern at any one point along the sensitive membrane. In either case, at low frequencies in the neighbourhood of which the half-cycle is longer than the quip time, so that most of the pressure gradients fall below the mean ($\frac{1}{4}$) of the whole distribution, the number of small pressure perturbations per quip is constant, and the probability density of the quips in the whole distribution is also constant. For this part of the distribution, therefore, the total number of pressure changes per narrow frequency band is constant and the curve relating total articulation with frequency will be linear if the frequency scale is also linear. The number of contributions and articulation content up to the top frequency in this part of the distribution will be proportional to the 4th power of that frequency.

At higher frequencies than this, where the pressure gradients are the result almost entirely of changes falling within the coherent and approximate Gaussian part of the distribution, the density of the quips in both time and space distributions will be approximately Gaussian in form. The total number of pressure changes passed by each filter therefore falls exponentially, and the curve relating total articulation with frequency will rise linearly if the frequency scale is logarithmic. The combination of these two sets of conditions is a linear characteristic, if the frequency scale is linear below the change-over frequency, and logarithmic above it. This is, of course, the well-known 'Koenig' scale published in the *Bell Record* for August 1949 (see also Figure 6).

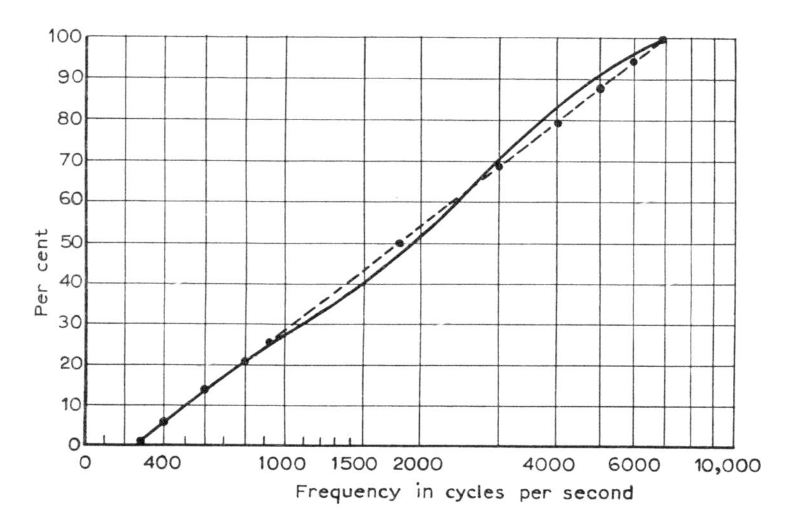

Fig. 6. The Koenig scale. Any two frequency bands showing equal percentage increments on this curve have the same inherent importance to articulation. Taken from fig. 5, *A New Frequency Scale for Acoustic Measurements*. W. Koenig, Bell Lab. Record, Vol. 27, No. 8. August, 1959. Points – – • – – – have been calculated from neogaussian statistics and a change-over frequency of 915 c/s.

Using the same normalisation constants as before,

(a) The 50% articulation point will occur at 1830 c/s.

(b) The mean of the whole distribution (the point of zero coherent deviation) and the change-over from linear frequency scale to logarithmic will occur at $\frac{1}{4}$ (25% articulation). This should occur at a frequency of log $\frac{1}{4}$/log $\frac{1}{4}$ $(25/50)^2$, or a half of that at the 50% articulation point. The 25% articulation point of change-over is therefore at 1830/2 or 915 c/s.

(c) The minimum articulation will fall at a frequency below which the total articulation (probability) is the indeterminacy 1·095%. This occurs at $\sqrt[4]{0{\cdot}01095}$ or 0·3236 times the frequency at the mean (915 c/s), i.e. at 296 c/s.

(d) The maximum articulation at the maximum coherent deviation above which there is no recognisable pattern, will occur at a frequency log $\frac{1}{4}(0{\cdot}01095)^2/\log\frac{1}{4}$ times the change-over frequency (915 c/s), or 6,880 c/s, at which point the total articulation from all lower frequencies will be 98·45%.

These theoretical characteristics have been plotted, together with Koenig's measured characteristics, in Figure 6. It will be seen that there is agreement between theory and measurements to within about 5% over the whole band. Koenig has suggested that the change-over frequency should be standardised at 1,000 c/s, though his reason for doing so does not seem to be clear.

We can easily extend the analysis to include speech intensities (each value of which will here be regarded as an equiprobable subdistribution of perturbations) usually expressed as pressures above atmospheric pressure. For this we shall take as normalising value a level of +70 dB, which is the average pressure within the syllable during the more recent series of articulation measurements reported by Fletcher & Galt in their paper published in March 1950. Since the effect of averaging all the levels within the syllable is largely to eliminate perturbations, this intensity level of +70 dB is what we have referred to as the colar pressure. All of the other statistical pressure levels may then be deduced as follows:

(a) The quip level will be below the colar level in the ratio $\sqrt{100/1{\cdot}095}=9{\cdot}556$, so that the quip level will be 70–19·6 or +50·4 dB.

(b) The average energy of the syllable as a whole will be $(1/P(R))^2$, or approximately 480^2 times that of the quip, which is equivalent to a level of +104 dB. From this it follows that the RMS syllable level will be close to +105·6 dB.

(c) The ratio of the average colar intensity to the maximum coherent level within the average syllable will be 1 to 5, so that, assuming a Neogaussian distribution, the maximum coherent level within the syllable should be 70 + 20 log 5 84 dB. Similarly the maximum intensity of the syllable as a whole will be +119·6 dB RMS. The dynamic range of speech has been found experimentally to be roughly 70 dB, and this gives +49·6 dB RMS for the minimum, which is thus reasonably in agreement with the quip level.

(d) Lastly, if we take speech at its minimum coherent pressure (quip) level of +50·4 dB, and further restrict it to the *time*-intervals within which the coherent pressure gradient is the minimum (in the quip at the mean of the time distribution in the neighbourhood of 915 c/s), the lowest detectable pressure perturbation will then be at a level of 50·4 dB+20 log P(R) or —3·2 dB. This is the level during a half-cycle of disturbance (pressure *rise*), so that the minimum detectable complete disturbance cycle level in the neighbourhood of 915 c/s should theoretically be about —0·2 dB. In fact our zero level is defined as this for 1,000 cycles per second pure tone. This suggests that the loudness of pure tone is also a matter of the frequency of occurrence of small gradients within it, its special pattern determining only its 'quality'.

As far as is known, all the above theoretical figures are within 1 dB of the values derived from actual measurements, so that deductions based on the assumption of Neogaussian statistics could give a valuable explanation of the statistics of actual speech under the conditions of articulation tests. The Neogaussian distribution might thus provide an analytical tool of great value in such problems as the design of communication systems to meet specified articulation standards, in speech recognition by machine, in the provision of a hypothetical speech test signal for various types of sound reproduction and communication systems, etc. The fact that articulation is basically statistical in form has been widely accepted for many decades now. Close accord between the articulation expected from various speech frequency components, assuming a Neogaussian distribution for the speech perturbations, and actual articulation measurements on speech is a clear indication that, under standardised speech conditions, the frequency of occurrence of small pressure changes is closely Neogaussian in form.

It is of course obvious that with some tens of thousands of sensitive nerves communicating with the brain of the listener, it is quite impossible for him to accept and use all of the detailed information available from all of them. The utmost he can do is to recognise and make use of the patterns of activated nerves which the incoming speech-waves set up in them. If this is the case for speech, it is to be expected also for sight, which is a sense which depends upon the light reaching the eye within which the number of sensitive nerves runs into hundreds of thousands. Quite apart from the fact that there is the same brain at the receiving end of the nervous system in the two cases, there are the following indications that Neogaussian statistics might be equally effective in any analysis of the light perturbations incident upon our eyes.

(a) It has been found that we can detect patterns in incoming light with approximately five times the sensitivity that we can apply to the identification of one single disturbance. For example, we can detect in a field of black dots on a white background a line of dots with an acuity which is about five times that with which we can separate a single dot from its neighbour.

(b) If we allow a long enough time in very low lighting conditions substantially to adapt ourselves to the dark, the result is generally that we attain approximately 200,000 times our sensitivity in ordinary daylight. In a Neogaussian distribution, we would expect this figure to be $(1/0{\cdot}002084)^2$, or a 230,000 increase.

(c) In a complex well-lighted picture, such as is available from a modern television set, it has been found experimentally possible to observe a pattern interference down to about 54 dB below the highlight intensity level in the picture itself. The figure to be expected from Neogaussian statistics is 53·6 dB (20 log 1/P(R)).

(d) For random interference below the general level of the picture, the acceptable interference level is about 31 dB below the double amplitude peak level of the picture. This is an average level of picture to noise of about 20 dB, a figure which should be compared with 19·6 dB to be expected from this theory.

This pattern recognition in our senses of sight and hearing is, of course, subconscious, and presumably the result of evolution in a continually changing natural environment in which the predominant fields of disturbances, and the stimulations as a result of which we are aware of them, can be 'explained' by similar statistical descriptions. It is apparently a mental characteristic of a great many forms of life on earth, presumably in all cases subconscious, except in man, by whom it can be consciously applied in sciences and technologies whenever analyses which neglect perturbations do not yield useful results. As an example of such a conscious use, let us now apply Neogaussian statistics to radio telecommunication through the atmosphere at microwave frequencies.

7

IN RADIO PROPAGATION AT MICROWAVE FREQUENCIES

> If a man begin with certainties, he shall end in doubts;
> but if he will be content to begin with doubts,
> he shall end in certainties.

Even though we now accept the fact that Neogaussian statistics often allow us accurately to account for the observed effects when qualitatively similar disturbances entering our eyes or ears come simultaneously from two different causal sets, it should not be concluded that *all* sight and sound information that reaches us can usefully be analysed in this way. It is probably true that we accept such information always through a pattern recognition process, but we can identify patterns by noting simultaneously several qualitative differences between a pattern and its background. When we do use more than one qualitative difference for this purpose, the presence of perturbations in the amplitudes of any one quality may have negligible effect on the recognition process. As an example, the printed word is often easily to be distinguished from the paper background on which it appears by its colour as well as by its intensity pattern, and the distinction between one printed letter and another is nearly always very large compared with slight differences between the same letter at various positions on the page. Small variations between different examples of the same letter, either in form or colour, do not normally limit the ease with which we can read. In general it appears that whenever we exercise control, either by selection of materials or environment or by the use of machines, we do so to reduce the effects of perturbations and relatively to increase those of the coherent components therefore. This makes it correspondingly easier for us to 'recognise' the latter and the patterning which derives from their coherence, and enables us more easily to 'understand' what is happening and formulate laws to 'explain' our observations. When we use our eyes or ears to extract the *maximum* amount of information possible from the many thousands of stimulated nerves which they contain, or when we increase our natural sensitivities by interposing telescopes, microphones, amplifiers, or other instrumentation, it will often follow that a complete analysis of our observations will involve not only the full coherent range of observation but will also extend through a change-over point into perturbations. In such cases the usual fundamental assumptions underlying any *rigorous* mathematical analysis become untenable, and we have, therefore, to revert to a statistical approach involving two causal sets, one for each domain to either 'side' of the change-over point.

By far the greater part of modern 'science' and the rigorous mathematics on which it is based is concerned with circumstances in which the perturbations have deliberately been made negligible, so that one, sometimes very complex, law can be used accurately to describe the whole field of changes of interest. In such cases we may usefully measure and calculate, assuming that each quantitative change will have one precise 'single value' at any one time. We may then use single-valued parameters, and associate the important ones together in one rigorous law without serious error. Even in this modern machine age, however, this is not always the case, and it may be instructive to take, for our last example of the application of doubly indeterminate statistics, a 'turbulent' section of the art of telecommunication, often regarded as one of the most advanced 'sciences' of our age. Whenever the transmission medium and terminating conditions can be controlled, progress in telecommunications has been astonishingly rapid. The machines devised to maintain coherent signals large compared with the perturbations or noise—satellites, repeaters, cables, transmitters, telephone, telegraph, television and sound broadcasting sets—are among the most sophisticated and widely used devices in the modern world. There is, however, another area of the telecommunication field in which we often have little or no choice but to transmit radio waves through the lower atmosphere, which is of course a turbulent medium so far quite uncontrollable. Whenever we do so over distances which result in important perturbations (fading) in the signal transmitted, our 'science' often fails to explain what actually occurs, and our technology is often inadequate to enable us to set up optimum transmission conditions. For the benefit of those who may not be aware of this regrettable state of affairs, here is some supporting evidence.

There are two 'Classical' propagation formulae in general use in radio system planning. These are the Fresnel Zone formula, used in estimating the necessary clearance over obstacles for 'free-space' conditions, and the Smooth Earth Diffraction formula for calculating the loss introduced by the earth and other obstacles when there is insufficient clearance. The first of these formulae ignores the effects of the atmosphere, but purports to allow for the interference effects between two parts of a radio field or 'rays' considered to have traversed paths of different lengths and/or suffered reflection somewhere along the route. In Fresnel Zone formulae, it is assumed that the effects of refractive index and velocity of propagation changes in the atmosphere are nearly the same along both paths, differences being negligible compared with the path-length difference and with the half-wavelength delay normally to be expected when reflection occurs intermediately along one path at a solid or liquid surface. Such an assumption was, of course, sooner or later bound to fail as wavelengths were reduced; its validity today at microwave frequencies may be estimated from the following quotation taken from Bell Telephone sources:

> 'The required clearance depends on the probability of inverse bending of the radio waves, and the amount necessary for satisfactory transmissions should increase more nearly as a square of the path length rather than according to the first Fresnel Zone rule, which varies as the square root of the distance.'

The classical diffraction formula assumes that the effects of an obstacle are related to its conducting properties. The validity of the classical diffraction formula for microwaves can be assessed from the following quotations, which come also from Bell Telephone sources:

> 'Thus it appears that the median field intensity is generally lower at short distances and higher at long distances than is predicted by the smooth earth theory.'
>
> 'The median signal levels . . . are hundreds of dB stronger than the value predicted by the classical theory based on a smooth spherical earth with a standard atmosphere.'

'The assumption of another value of K would not provide a better fit with the experimental data . . . there is no apparent method of adjusting the smooth earth theory by changes in the values of K or the ground constants to fit the available experimental data.'

'This means that theoretical curves based on the smooth earth theory may be in serious error at distances beyond 100 miles for 30 to 40 Mc, beyond 60 miles for 300 Mc, and beyond 30 miles for 3,000 Mc. The apparent failure of the smooth earth theory is most noticeable in the region where it should be most accurate.'

The effect of the atmosphere on the propagation of radio waves can be attributed to small changes in the refractive index, which is usually given by the formula:

$$N=(n-1)\,10^{6}=\frac{77\cdot6}{T}\left(p+4{,}810\,\frac{e}{T}\right)$$

Where p = total pressure in millibars

e = partial pressure of water vapour in millibars

T = absolute temperature = °C + 273.

n = refractive index.

N, the difference between the refractive index and unity expressed in millionths, has been called the Refractivity of the air.

This expression is considered to be good to 0·5% in N for frequencies up to 30,000 Mc/s, and over normally encountered ranges of temperature, pressure, and humidity. In practice the refractive index varies between unity and about 1·0004, and the refractivity has an average figure of about 300. Radio engineers have standardised a 'Standard Radio Atmosphere' as one in which radio waves are propagated along a path which has an average curvature a quarter of that of idealised smooth earth (assumed to be a sphere of radius 3,960 miles and to have, therefore, a surface curvature one curth; see Chapter 4). In such an atmosphere the relative position of a radio beam directed horizontally and smooth earth may be calculated by regarding the beam as travelling along a straight path (zero curvature) above a theoretically smooth earth the surface curvature of which is three-quarters of that of the surface of a sphere of radius 3,960 miles, that is to say, the radio path may be regarded as straight above an earth, whose radius is four-thirds of 3,960 miles. This is often expressed by saying that the 'Effective Earth's Radius Factor' for the standard radio atmosphere is K=4/3. This standard radio atmosphere has theoretically a refractivity gradient in the vertical profile of 12 N units per 1,000 ft. of height, and in it the path of a radio wave has a curvature of 0·25 curth, so that the curvature difference between the path of the radio beam and idealised smooth earth is $\frac{1}{4}-1$, or $-3/4$ curth.

It is of course tacitly assumed in the above formula for the refractive index N that the relevant thermal conditions in any given volume of air are precisely defined by its temperature T. On the other hand, it was shown in Chapter 4 that actual distributions of 'temperatures' in the lower atmosphere can be explained much more satisfactorily by assuming that they are the result of two causal sets, one of which *alone* would lead to isotropic and random perturbations in a particular small volume, the whole of which would therefore be identified by a single temperature, the other leading to the longer-term and larger-scale temperature *differences* and gradients in the vertical profile. In these circumstances there will be two approximate statistical laws determining the temperature distributions in the lower atmosphere, one of them mostly associated with the perturbations within parts of a particular small height intercept over which the temperature is constant, and the other mostly appropriate to the temperature changes and gradients which appear across larger height intercepts during longer time-intervals. Since both causal sets to some extent affect the turbulence within volumes allegedly at a single temperature, conditions such as this cannot be 'explained' by one rigorous formula. Applying the basic ideas of Chapter 4, we should treat each 'single temperature' as an equiprobable distribution of disturbances within small parts of the zone height intercept (approximately 22·7 ft.), a few of the fluctuations among them adding coherently to form more than one 'temperature' across height intercepts greater than about 22·7 ft. and during time-intervals of more than about three minutes. The effects of the atmosphere on radio propagation, therefore, can be approximated by one set of propagation laws for radio wavelengths more than about 22·7 ft. (radio frequencies less than 43 Mc/s) and another different set of laws for shorter wavelengths (frequencies well above 43 Mc/s).

Here we shall be discussing propagation at microwave frequencies well above 43 Mc/s, the radio waves being launched in approximately a horizontal direction. We shall briefly comment on other conditions later, but it will, in the first instance, be assumed that conditions are those to which we have already referred as SATRA (Standard, Average Turbulent, Radio Atmosphere). These conditions permit the direct application of the normalised Neogaussian solution for the lower atmosphere derived in Chapter 4, which is:

The *Zone Height* intercept, within which any given very small temperature perturbation is equally probably a gravitational disturbance associated with the solar tide as a diurnal cycle temperature change rising from the ground below, is 22·71 ft.

The *Maximum Coherent Deviation* of the bending of the path of radio waves is that which is along the assumed horizontal layering, that is to say, it is 1 curth.
(This is the normalisation constant of the distribution in space)

The *Time Length of the Coherent Distribution* of solar cycle temperature rises is 1 day.
(This is the normalisation constant of the distribution in time)

The *Coherence Limit* (S %) is 98·905 %.

The *Indeterminacy* (100-S) % is 1·095 %.

The *Standard Deviation* (σ) of the Distribution of Zones is 0·4363 curth.

The *Basic Uncertainty* of the whole distribution of path curvatures $P(R)=\sigma^2 \times (100\text{-}S)/100$ is 0·002084 curth.

In this solution SATRA conditions have been assumed, i.e.

(a) The air is dry, so that there are no change of state of water and latent heat effects involved. This is often true overland, but often untrue over water.

(b) The atmosphere is layered horizontally, so that the layers and isopleths of temperature have an average curvature of 1 curth. This is generally true over nearly flat terrain.

(c) The diurnal and annual temperature cycles have their average intensities, which are those at latitude $37\frac{1}{2}°$ at the equinoxes.

We can at once refer to two experimental facts which suggest strongly that the statistics of radio propagation of microwaves are closely Neogaussian in form. The first of these is that it is well known that the mean bending in the atmosphere is 0·25 curth, which is of course also the mean of our Neogaussian distribution.

Secondly, it has been experimentally established (see *Proceedings of the I.R.E.*, October 1955, Bean & Meaney) that:

'One of the major conclusions is that the observed dependence of transmission loss beyond the horizon upon N is five times greater than that indicated by standard propagation theory.'

This ratio of the maximum coherent variate (in this case path bending) to mean of 5 to 1 (see Figure 1) is perhaps the most prominent feature of the Neogaussian distribution.

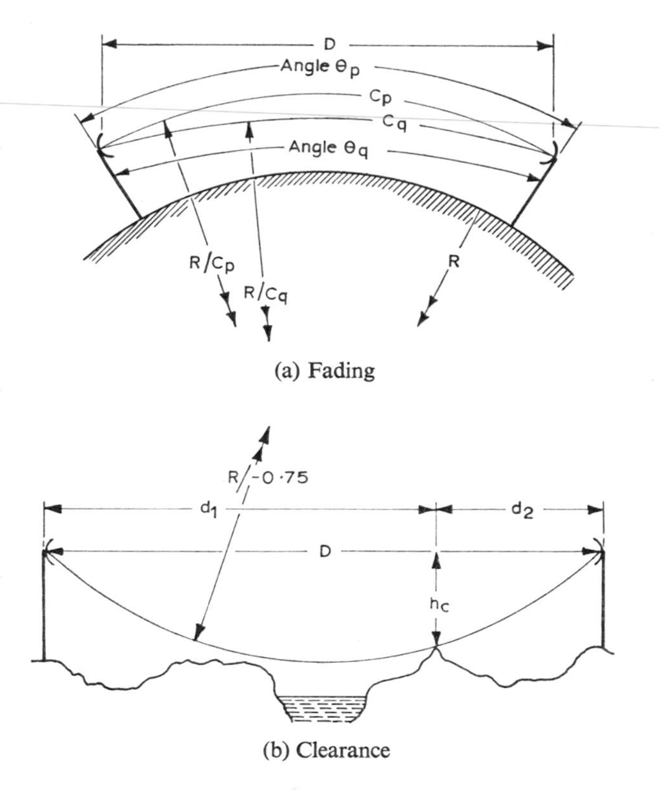

Fig. 7. Line-of-sight paths

Line-of-sight Propagation (fading). Consider the two circular paths above smooth earth of curvature Cp and Cq in Figure 7. It can be shown that the approximate difference in the path lengths along the two paths Cp and Cq is given by

$$\Delta D = \frac{D^3}{24R^2}(C_p{}^2 - C_q{}^2)$$

where D is the path length in miles, R is the radius of the earth (assumed to be 3,960 miles), and Cp and Cq are the curvatures of the two paths in curths. The angular path difference in radians for a radio frequency wave (F Mc/s) travelling along these paths is therefore

$$0{\cdot}0896 \times 10^{-6}\ D^3\ F(C_p{}^2 - C_q{}^2)$$

In this formula, the only term associated with propagation conditions in the atmosphere is $(C_p{}^2 - C_q{}^2)$, which will accordingly be called the Angular Path Difference Factor.

In line-of-sight paths, and particularly for grazing paths over rough ground, some contribution to turbulence from wind and other random factors relatively uncorrelated with the diurnal cycle is to be expected close to the ground. The more there is of this, the greater will be the mixing and the height intercept over which the mean bending is almost constant, and the less will be the chance that there will be cancellations due to different path curvatures occurring simultaneously within the height intercept spanned by a receiving antenna aperture. 'Multipath' effects generally are much less severe in winter and over rough ground than in summer and during still weather. The statistical theory provides a method of estimating the fading characteristics over line-of-sight paths only in conditions in which the turbulence is a minimum and the result of the solar cycle and gravitational perturbations alone. The theory is therefore useful for planning high-quality microwave systems in still weather, and unreliable (often pessimistic) for estimating the average fading depth throughout the year over a line-of-sight path.

Two 'weather' conditions will be considered. In one of them, either because at the time the rate of change of temperature in the diurnal cycle is very small, or because the weather is, for other reasons, sufficiently turbulent at the time considered, the mean path bending is constant over a height equal to or greater than the aperture heights of both antennas. Obviously, in this case, the received field will be substantially random about a median which varies slowly with the diurnal cycle but is always nearly the same over the dish *as a whole*. The received signal will then be substantially a Rayleigh one. As is well known, its instantaneous amplitudes will fall below its root-mean-square by 10 dB for a maximum of 10% of the time, by 20 dB for 1% of the time, by

30 dB for a maximum of 0·1% of the time, and so on, as shown in Figure 8. In the other weather conditions, it will be assumed that the turbulence is nearly the minimum during the diurnal cycle, and is such that there are just *two layers with different path* curvatures within the aperture height. This is the worst possible case. When it occurs, the systematic components of the field associated with each of the two path curvatures may completely cancel one another, leaving only the very small random field components, perhaps about 54 dB (20 log P(R)) lower in level, since their probability of occurrence averaged over the diurnal cycle might be only about 0·002084. Slow fading under these conditions would therefore be much more severe than is likely under conditions in which there is only Rayleigh fading, and there seems an obvious case for limiting the path length to make such almost complete cancellation impossible, or at any rate improbable, during S% of the time.

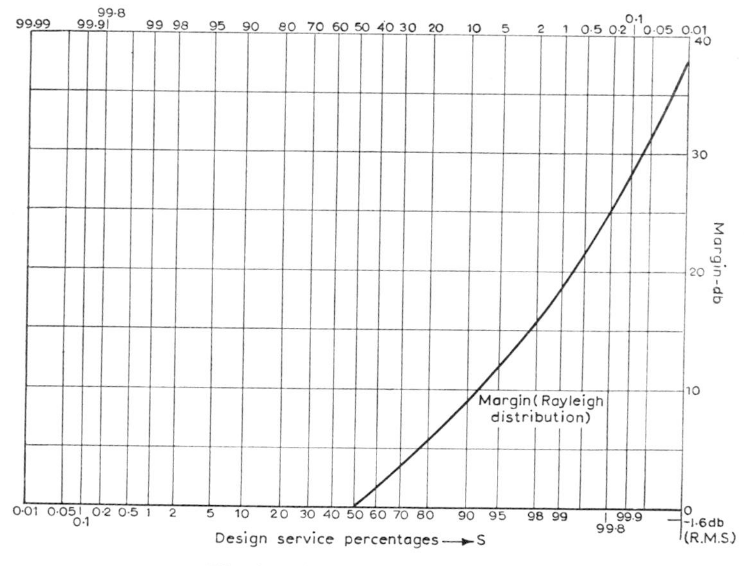

Fig. 8. The Rayleigh distribution

With this in mind the main facts of importance in 'designing' a line-of-sight single section or link of a chain of such links are:

(a) For a given frequency the path length should be as long as possible consistent with acceptable outages, even during minimum turbulence, for practical margins during the design service percentage of time (S%). The margin is the maximum possible increase in sensitivity of a receiver above that of its normal 'no fading' setting without objectionable deterioration of the signal in its output.

(b) Outage will be much less with a given margin if it occurs only as a result of Rayleigh fading.

(c) The maximum angular path difference factor during layering can be estimated from the fact that the layer spans a height $\sqrt{100/(100\text{-}S)}$ times that of the zone, and therefore the standard deviation of layer heights is $\sqrt[4]{(100\text{-}S)/100}$ times that of zones. As S is 98·905% and $\sigma = 0{\cdot}4363$, the standard deviation of layers is $\sigma \times \sqrt[4]{(100\text{-}S)/100}$ or 0·141, the maximum layer path curvature deviation in 98·905% of the time is $\sqrt[4]{(100\text{-}S)/100} \times 1{\cdot}0 = 0{\cdot}323$ and the maximum layer curvature in 98·9% of the time is then $0{\cdot}25 + 0{\cdot}323 = 0{\cdot}573$ curth, if the mean curvature $C_a = 0{\cdot}25$.

The angular path difference is obviously a maximum when the path curvature in one part of the dish has this maximum value of 0·573, while that in the other half is zero (($C_p^2 - C_q^2$) is a maximum when Cq is 0), and when this is the case the greatest angular path difference 'factor' is $0{\cdot}573^2 - 0^2 = 0{\cdot}33$.

There seems an obvious case for controlling the maximum path difference in S% of the time so that under layering conditions it is always a little less than π radians. There should then be no outage resulting from the relatively slow cancellations between layers (Gaussian fading) and only Rayleigh fading, which can be reduced effectively by the much smaller margins necessary for this type of fading. The maximum path difference during layering should be kept below π radians by an amount such that the resultant drop in level due to each component of the fading is not more than half the available total system margin. This allows half of the margin for the Rayleigh fading during high turbulence and the shorter time-intervals associated with zones, and half for the drop in level of the mean of the Rayleigh fading due to partial cancellation between layers during their longer time-intervals. As the total available system margin is generally between 20 and 40 dB, the angular path difference during layering should be kept to a maximum a little less than π radians, say about 3 radians.

Hence,

$$0{\cdot}0896\times10^{-6}\times D^3\times F(C_a+0{\cdot}323)^2=3{\cdot}0$$

$$D\times\sqrt[3]{F}=322\,(C_a+0{\cdot}323)^{-2/3}.$$

If $C_a=0{\cdot}25$, $D\times\sqrt[3]{F}=470$, where, as before, D is the path length in miles, F is the radio frequency in Mc/s.

This formula leads to results which are in close agreement with experimentally determined 'good practice' for land paths over a a very wide range of frequencies. In it, it has been assumed that the average curvature in the atmosphere above the ground is the mean of the whole distribution 0·25 curth. Over water it may frequently happen that there is a stratum near the water's surface in which practically all of the diurnal cycle heat is being carried upwards as latent heat to be released at a greater height above ground. In this case conditions in the complete layer of atmosphere, including the height intercept in which the latent heat is released, may at times be such that the total amount of heating from below is almost double what would otherwise be the case, so that the mean path curvature could be nearly 0·5 curth ($K=2$) instead of 0·25 curth. If we assume an extreme effective earth's radius factor $K=1{\cdot}9$ the maximum path length for a line-of-sight path, to avoid Gaussian fading between layers, becomes $-D\times\sqrt[3]{F}=375$ for paths above or near water surfaces.

Similarly, the theoretical extreme minimum path bending in any layer above the ground in part of which the diurnal cycle heating is mostly absorbed during evaporation of water in the air, will be a half of the 0·25 curth mean of the whole distribution in dry air, that is to say, it will be 0·125 curth. In practice, paths over deserts do not usually fall below a mean curvature of about 0·2 curth, and for this case $D\times\sqrt[3]{F}=500$. The equation $D^3\sqrt{F}=470$ applies in SATRA conditions.

The above equations define the path lengths for Rayleigh fading conditions, and they give the basic relationship for the control of multipath fading by avoiding the slow near-cancellations associated with 'Gaussian' fading at any rate for S% (approximately 99%) of the time.

It should be emphasised that the need to limit path lengths in this way arises because the antenna aperture heights used in practical microwave installations are necessarily at present much greater than the height intercept over which the dispersion of the radiation in the anisotropic air turbulence results, effectively, in a single path and Rayleigh fading. The height intercept traversed by a radio beam along a path D miles long, in SATRA conditions (effective earth's radius factor $K=4/3$), averages:

$$D^2 \times 5{,}280 \times 3/4 \times 2 \times 3{,}960 = D^2/2 \text{ ft.}$$

so that during about 99% of very long periods of time, random path bending and Rayleigh fading will occur over a height intercept $D^2 \times 0{\cdot}002084/2$, or approximately $D^2/1{,}000$ ft. It follows from this that, in SATRA conditions, multipath layering and Gaussian deep fading is to be expected if the vertical dimension of the antenna aperture exceeds 0·23 ft. for a 15-mile path, or 3·6 ft. for a 60-mile path. Practical antenna sizes are normally always much larger than this, 10 ft. dishes being quite usual.

Line-of-sight paths, clearance requirements

Figure 7 (b) has been drawn assuming flat earth, a layer of the minimum (zero) coherent path curvature deviation near the ground, and a mean path of radiation which results from the random turbulence of $-3/4$ curth relative to flat earth (SATRA). From a theorem of intersecting chords in a circle, $d_1\ d_2 = h_c \times 2R/-0{\cdot}75$ whence h_c (the required clearance above earth)

$$= 3d_1\ d_2/8R \text{ miles}$$
$$= d_1\ d_2/2 \text{ ft.}$$

If the predominant obstruction happens to be at or near the centre of the path, this clearance requirement becomes $D^2/8$, where D is the path length in miles. The required clearance thus varies as the square of the path length, in accord with the recommended best practice taken from Bell System sources and quoted at the beginning of this chapter. It may, of course, be considered advisable to add a further 10 or 20 ft. of clearance to allow for the growth of vegetation and for other possible small changes on the surface of the earth.

The theory as it applies to line-of-sight paths may be elaborated to provide estimates of the fading to be expected when paths are too long or the clearance is less than indicated using the formula given above. It is also possible to extend it to provide estimates of the band-widths available in multi-channel microwave systems. In these and in many other matters there is very satisfactory agreement between theoretical conclusions and good practice recommendations built up as a result of many years of experience in many parts of the world. Perhaps the one innovation, which, it appears from theory, might result in improved propagation conditions, is the obvious one that, in certain rather special cases, in still weather and over relatively smooth earth, it might be advantageous to use antennas of which the apertures are considerably greater horizontally than in the vertical direction. This does not seem to have been tried out in practice so far.

Transmission losses over longer paths

At the radio frequencies in general use for microwave systems (below about 11,000 Mc/s) absorption by the air may be neglected, and the transmission losses encountered during propagation between the antennas of a line-of-sight path involve only two items:

The fraction of the total energy radiated in the direction of the receiver and intercepted by the receiving aerial, neglecting all effects of the atmosphere. This is the Free Space Loss. As it depends only on the path length and the terminal equipment, it is not truly a propagation term, and will not be considered here.

An allowance for the effects of the atmosphere on the received signal, i.e. for the fading. It is usual to provide for this by planning so that there is sufficient clearance over obstacles in nearly all atmospheric path bending conditions, and then designing equipment so that it has sufficient margin to maintain satisfactory transmission standards for an acceptably large percentage of the time.

We have already indicated how the clearance above obstacles and the path length may be proportioned so as to control the fading over line-of-sight paths. If a transmitting antenna is erected and set so that it radiates mostly horizontally and in the direction of the receiver, we shall find that, on the average in dry air, half of the main lobe of the radiation is cut off by the ground, at a distance well beyond the line-of-sight distance, which we shall here define as the radio horizon distance. From this definition, of course, a receiving antenna on the ground at the radio horizon can accept only half the radiated field, and it follows that radio transmission over still longer paths will involve a loss at the radio horizon of 6 dB. We shall here associate the radio horizon by definition with a mean curvature of path bending of 0·25 curth, (K=4/3).

Beyond the radio horizon, that half of the beam which is not intercepted by the ground will rise through the lower atmosphere, encountering zones in which there are equiprobable and nearly isotropic temperature perturbations, although each zone is associated with one single 'value' of temperature. The average bending in this turbulence in dry air which results from transmission through many successive zones along a radio path is 0·25 curth. Although the perturbations within zones are substantially equiprobable and isotropic, the air in adjacent intercepts in the vertical profile, when grouped into heights (layers) which include many zones, will contain coherent temperature *differences*, so that there will be across such larger height intercepts, coherent path bending and temperature gradients which correlate with the diurnal cycle rising from the ground below (see Figure 2). Whenever the coherent bending in such layers is greater than the mean (generally 0·25 curth), there will be some contribution to the field receivable at ground-level beyond the radio horizon calculated on the assumption of a mean bending of 0·25 curth. The coherent signal level in an antenna on the ground at a point beyond the radio horizon is determined by the frequency of occurrence of the bending greater than the mean necessary to reach the point concerned. Since the frequency of occurrence of bending greater than the mean in the Neogaussian distribution falls exponentially with the deviation, the loss in logarithmic units (decibels) to the mostly coherent signal near the ground in the diffraction section of a long path will be approximately proportional to the distance beyond the radio horizon. A formula of this type has, for rather different reasons, been used and found experimentally to be satisfactory for the calculation of the diffraction loss in a layered troposphere above a relatively smooth spherical obstacle such as the earth.

Fading in the diffraction section of a long path just beyond the radio horizon would be expected to be greater than for line-of-sight paths because of the statistical nature of the obstacle effect of the earth in various atmospheric bending conditions, and the increasing randomness in the distribution as the bending deviation increases and its frequency of occurrence falls towards that of the perturbations within a single 'temperature'. From the very nature of the Neogaussian distribution, however, we should always expect a small, approximately Rayleigh distributed, random field component in all bending conditions and for all values of effective earth's radius factor K which, in dry air and over rough but approximately flat terrain, comes from the 1·09% indeterminacy in the probability of the bending. This is an essential feature of the Neogaussian distribution. It leads to a random field with an average value, over the whole distribution of path curvatures, corresponding to the basic uncertainty P(R)=0·002084, i.e. at a level of 53·6 dB below that of that half of the coherent beam rising through the air beyond the radio horizon (SATRA conditions).

Under the standardised conditions of Chapter 4 (SATRA), the end of the diffraction section beyond the horizon may now be calculated as follows:

(a) Under our particular idealised conditions the loss beyond the radio horizon at the change-over point at the end of the diffraction section will be 53·6 dB. This is, of course, additional to the 6 dB radio horizon loss.

(b) At the end of the diffraction section, and centred on a total maximum bending of 1·25 curth, of which 1 curth is the path bending and 0·25 curth the mean bending in the turbulence in the zones along that path, there will be a sub-distribution of curvature perturbations extending from 1·0 to 1·5 (see Figure 1). At the end of the diffraction section this curvature range embraces a sub-distribution of perturbations randomly distributed over one complete wavelength of the radiation. There is thus a sub-distribution of small contributions to the received signal which are substantially in random phase over an angular path length difference of 360° (2 π radians). Hence, at the end of the diffraction section,

$$0{\cdot}0896\times10^{-6}\,D^3\,F\,(1{\cdot}5^2-1{\cdot}0^2)=2\,\pi$$

$$D^3\sqrt{F}=383.$$

Combining these two conclusions, diffraction under the standardised conditions we have postulated will end at a loss beyond the radio horizon of 53·6 dB, at a distance $383/{}^3\sqrt{F}$ miles, the average loss per mile in the diffraction section therefore being $53{\cdot}6\,{}^3\sqrt{F}/383$ or $0{\cdot}14\,{}^3\sqrt{F}$ dB per mile. Beyond this diffraction section it is to be expected that there would be a change-over to an approximately Rayleigh distributed random or scatter field. That this is so has been most elegantly confirmed in tests carried out by the Navy Electronics Laboratory in San Diego, and published by L. G. Trolese in *Proceedings of the I.R.E.* for October 1955. Figure 5 of Trolese's paper has been reproduced as Figure 9 here.

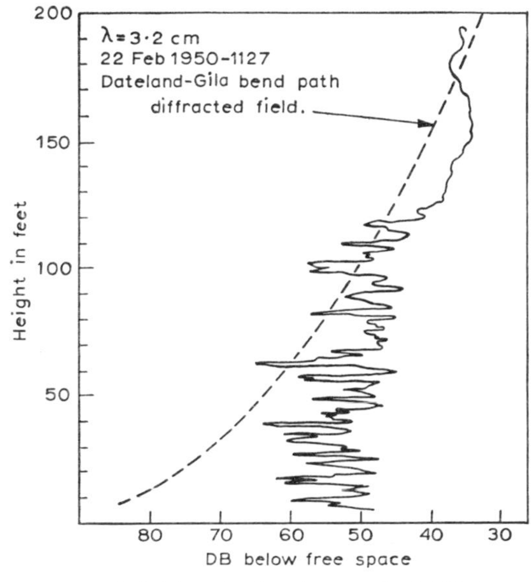

Fig. 9. The random field at the end of the diffraction section

Briefly to recapitulate, the air in the lower atmosphere has a basic temperature, by which is meant the temperature it would have in the absence of the diurnal solar cycle, due to the gravitational disturbances associated with the solar tide. The mean force holding the air to the surface of the earth and responsible for its horizontal layering is, of course, the gravitational attraction of the earth. The earth is, however, not a perfect sphere, and is rotating rapidly on its own axis during its orbit round the sun, and as a result the air is also subject to the important gravitational perturbations associated with the solar tide; these constitute its basic temperature. Near the ground, however, and to a progressively less extent at higher altitudes, the air is additionally heated above its basic temperature by the diurnal solar cycle as it rises slowly upwards in the turbulence from the ground below (see Figure 2). The air at the tropopause is approximately at its basic temperature; at very low altitudes in the troposphere, the air is on the average heated above its basic temperature by an amount which depends on the mean of the diurnal 'cycle' rising through the air, and the deviations from the mean during this cycle.

The coherent diurnal cycle of temperature rising in the vertical profile may be broken down into very small changes (fluctuations) the probability of which is no more than that of similar disturbances which result from the gravitational perturbations responsible for the basic temperature. On the average during the diurnal cycle, these equiprobable sub-distributions extend over height intercepts of about 22·7 ft. and persist for time-intervals of about three minutes. Simply as a result of the coherent vertical component in the turbulence, the average bendings in many successive zones add to form a curvature a quarter of that of the curvature of the isopleths of temperature, which we have here assumed are horizontal, i.e. parallel to smooth earth. This mean bending is the inevitable consequence of the anisotropic turbulence in a horizontally layered atmosphere additionally heated from below.

Just beyond the radio horizon where there is a loss of 6 dB, the average obstacle effect of the earth, the signal level depends on the frequency of occurrence of the bending necessary for propagation to the point concerned, i.e. to the position of the effective radio horizon for path bending in the air greater than 0·25 curth. Since the frequencies of occurrence in the Neogaussian distribution fall exponentially with deviation, the loss in decibels per mile in the diffraction section will be approximately constant. At a sufficiently great distance, however, the probability of the necessary coherent bending in all the zones falls to that of the random perturbations which are present throughout the whole of the lower atmosphere as the result of the gravitational disturbances responsible for its basic temperature, and when this occurs there is a change-over to a Rayleigh distributed 'random' or 'scatter' signal.

During the last two or three decades, theories to account for this phenomenon have been based on multiple reflections in the air, on reflection from the tropopause, on contributions from a particular scattering volume at great heights above the ground, and so forth. All too frequently it is forgotten that rigorous mathematical formulae are at best only approximations. All too rarely is there any attempt to identify the point at which the over-simplified basic assumptions underlying the mathematics become untenable. Diffraction calculations of propagation through the troposphere have been a typical case of such an omission. Considered in small enough height intercepts the atmosphere is made up of 'single temperature' sub-volumes (zones) within which there are only random path bending perturbations. At the distance where the diffraction loss beyond the horizon is large enough, so that the probability of the field there is small enough, the received field changes (with the statistics) to a scatter or Rayleigh distribution.

At the change-over point, the coherent radio wave-front disintegrates, and beyond this point identification of 'rays' becomes impossible. The radiation is then of the same general form as the light in an evenly illuminated shadowless room. Thereafter we may regard it simply as a random or scatter field occupying most of the space below the coherent beam, with a median direction of transmission but with indefinitely large numbers of small random path curvatures equiprobably distributed over a finite curvature range about this median. The change-over point to which we have so far referred is the one, of course, at ground-level. The field intensity at points vertically above it must be the result of radiation traversing zones along paths from the radio horizon with mean curvatures which become progressively less than 1·0 curth with increasing height, and the field therefore becomes increasingly coherent with height. The zone densities, coherence, and intensity of the radiation all increase with height above ground, and there is thus a miniature diffraction section in the vertical profile just below the mean height of the lower 'edge' of the unobstructed half of the beam.

Over the horizon losses to the random field

It is a comparatively simple matter to estimate the loss per mile to the random field beyond the change-over point. As the beam rises further and further into the troposphere along an average path which has a curvature relative to smooth earth of −0·75 curth, the maximum coherent path bending deviation in the curvature distribution will be through the zones along the horizontal layering, the median direction of the random field from these zones having a downward bending relative to the horizontal path of 0·25 curth. There is therefore a random or scatter field propagated as a whole downward toward the earth, where it may be received in an antenna on the ground as a substantially Rayleigh distributed 'random' or 'scatter' signal, at any rate during a short time-interval and in an antenna spanning a relatively small finite height intercept. Up to the change-over point, the field at ground level is predominantly coherent, and the losses are due to the earth obstruction to a coherent wave propagating mostly in a direction normal to its wave-front. The signal strength up to this point depends very much on the radio frequency used. Beyond the change-over point, the greater part of the received signal is the result of the small random path curvature perturbations in the equiprobable zones throughout the atmosphere along all coherent path curvatures up to the maximum deviation of 1·0 curth, so that at wavelengths much less than the zone height and in correspondingly short time-intervals, the loss at ground-level beyond the change-over point is largely independent of the radio frequency.

The distance (D_{co}) from radio horizon to change-over point already calculated applies to the change-over at ground-level. As already pointed out, a miniature diffraction section immediately above the change-over at ground-level extends in the vertical profile up to the average height of the lower edge of the unobstructed half of the beam beyond the radio horizon. In SATRA conditions this lower 'edge' has an average height $(D_{co})^2/2$ ft. above the change-over at ground-level. There are, of course, other change-over points at higher altitudes above ground, these lying along a line from the change-over point at ground-level with a curvature $-3/4$ curth (i.e. parallel to the coherent beam) relative to the earth below. In SATRA conditions the height within which there is one complete diurnal cycle of temperature changes rising in the turbulence is 10,900 ft. (see Chapter 4), and the unobstructed half of the beam will rise through this height of atmosphere over a distance at ground-level equal to $\sqrt{2\times 10{,}900}$, or 148 miles (D_{ti}). Once the loss along smooth earth exceeds 53·6 dB, the received field at ground-level will be substantially Rayleigh distributed, with a mean level at any one place which will vary in time depending on the zone densities in the distribution and the probability of the coherent temperature gradient deviations coming from the diurnal cycle. Its minimum in 98·905% of the slow fading will, of course, occur for 1·095% of the time, and at a distance of 148 miles beyond the change-over point, therefore, this minimum level of the Rayleigh distributed signal will be a further 10 log 0·01095 or 19·6 dB below the random field level at the change-over point. The loss per mile to the random field beyond the change-over point is thus 19·6/148, or 0·133 dB per mile. This is of course the loss to the mean level of a Rayleigh distribution when the latter is at its minimum in 98·905% of the slow fading. We have here ignored a small section of the route 10 miles or so long, immediately following the change-over at ground-level, within which the signal originates from the first zone of height or from any larger height intercept over which substantially equiprobable conditions are set up as a result of additional turbulence near rough ground. Over this 10 miles or so of path length, particularly in turbulent weather and at high radio frequencies, there should be substantially no loss to the minimum signal in about 99% of the time.

Beyond this section, the loss per mile to the Rayleigh distributed 'scatter' field at the minimum in about 99% of the slow fading will continue to be 0·133 dB per mile in SATRA conditions up to the distance above which the beam rises to the tropopause, since the only difference between the upper strata of the troposphere and the lowest diurnal cycle height intercept is that, at the greater heights, the coherent diurnal cycle is lost and there is therefore substantially a random distribution of temperature perturbations or zones in dry air without the coherence leading to the diurnal cycle-wave. In SATRA conditions this loss therefore continues up to $\sqrt{2\times 43{,}600}$, or approximately 300 miles beyond the change-over point.

Above the troposphere, and between it and the ionosphere about 47½ miles above the ground, there is a section of the atmosphere which is often called the 'stratosphere'. This includes a stratum about 10 miles thick within which the average temperature is approximately constant (the isosphere), so that the field strength at ground-level from the random field generated as the beam rises through this stratum will attenuate at approximately the same rate as from the troposphere. Over this section of a route, approximately 450 miles from the change-over point toward the radio horizon nearest the receiver, the level of the Rayleigh distributed signal at the minimum in the slow fading continues to fall at 0·133 dB per mile (in SATRA). Above the isosphere there is a still higher layer, the mesoincline, about 12 miles thick, within which the temperature rises almost to the same value as at ground-level, and this in turn is followed by a still higher stratum, the mesodecline, about 17 or 18 miles thick, within which the temperature falls again approximately to that in the isosphere. Above this is the ionosphere. It is believed that the rise of temperature to that at the mesopeak between the mesoincline and the mesodecline is due to the absorption of incoming radiation, particularly below 2,900 Å, by ozone. In the troposphere, convection is in control, the energy of the diurnal cycle being carried upwards in the turbulence mostly as a result of temperature changes coming from the earth below, at any rate above ground. Above the sea an appreciable part of this energy is released over a wide range of heights by condensation of water vapour carried up in the turbulence. The stratosphere, on the other hand, is mostly in radiation balance, the turbulence depending on the diurnal cycle and the ozone distribution. The amount of ozone in the stratosphere is relatively quite small, and its effects on the kinetic properties of the air are negligible. As a result of the turbulence this ozone is, however, distributed over quite a wide range of height, and the approximate linear temperature gradients in the mesoincline and the mesodecline are presumably the result of almost uniform distribution of the ozone by the turbulence. Based on these assumptions, it is possible to calculate statistically the losses per mile that will be produced, and these vary between about 0·08 dB per mile and about 0·17 dB per mile, with an average value over the whole distance up to 700 miles beyond the change-over of about 0·125 dB per mile. The beam beyond the horizon will rise to the base of the ionosphere at a height of nearly 48 miles above earth, over a distance of about 750 miles beyond the radio horizon. Contributions to the signal at ground-level from the ionosphere are quite different in character, and quite different frequencies and launching practices have to be used for transmissions over path lengths from 900 to 1,200 miles.

It is not the intention here to give the full statistical theory of propagation at microwave frequencies, but simply to show how Neogaussian statistics may be applied, with sufficient evidence to confirm that deductions from the statistical theory are in good accord with measurements. In a few instances, conclusions from the statistical theory run counter to the propagation rules at present frequently used in radio system planning, and it is therefore proposed now to discuss some of the characteristics of radio propagation at great distances beyond the horizon, with particular regard to areas in which previously held theories may not always have led to the best results.

Fading beyond the horizon

It is of course all-important that radio signals beyond the horizon should be measured over time-intervals and antenna aperture heights within which they conform to recognisable statistical sub-distributions. The more rapid fading in short time-intervals should be averaged over intervals of time within which it nearly always (S%) forms substantially Rayleigh sub-distributions. These sub-distributions will, as a result of the few coherent components within them, add over longer intervals of time to form the approximately Gaussian distributed slow fading, at any rate in dry air and over roughly level ground. From the statistical theory it follows that, in SATRA conditions, the signal should be averaged over no more than about three minutes for path lengths 148 miles beyond the change-over point, and for proportionally longer periods of time over the longer paths up to about 700 miles. There may obviously be periods of time during which Rayleigh fading conditions persist for much longer time-intervals than this, but a consistent analysis can be expected only when nearly all signal sub-distributions are substantially Rayleigh distributed, not just a great many of them. At 148 miles beyond the change-over point, the Rayleigh mean levels in the slow fading will rise and fall with the zone densities of the layers from which the random field happens to emanate, that is to say, the range between the mean level and the minimum in 98·9% of the time will, in SATRA conditions, be about 14·4 dB. This corresponds to a standard deviation of 6·3 dB at this distance. For longer paths the fading range and standard deviation will fall inversely proportional to the path length between radio horizons, the mean of the signal falling towards the minimum in about 99% of the time as already calculated above. The fading range at the equinoxes for path lengths about 650 miles between radio horizons is therefore expected to be only about 3·6 dB (SATRA), though of course the absolute loss at this great distance will be very high. By far the most predictable level in the received signal is the minimum in about 99% of the time, and a great deal of avoidable indeterminacy is introduced when data are compiled in terms of average values in the slow fading.

The behaviour of antennas in the random field

The full slow fading range of the received signal will not be recorded if signals are averaged over intervals of time within which there may be appreciable departures from the Rayleigh distribution. For much the same reason, interference effects are likely in any antenna the aperture height of which spans a curvature range larger than that within which there is at any one time an equiprobable distribution. At the minimum of the slow fading in 99% of the time approximately, therefore, and in SATRA conditions, the antenna aperture height should be such that it spans a curvature range not greater than 0·00208 curth. This is equivalent to saying that there may be different path curvatures and partial cancellation of the slowly fading mean signals within any antenna the aperture height of which is greater than $D^2/$1,000 ft. approximately, where D is the distance between radio horizons in miles. The application of this rule shows that, for part of the time there may be cancellation and apparent 'loss-of-antenna-gain' if the antennas are larger than 10 ft. at 100 miles between radio horizons, or 90 ft. at 300 miles between radio horizons. Tests carried out in Arizona have demonstrated loss of antenna gain in 4 ft. dishes over a 50-mile path, and other tests in the United States have demonstrated no such effects in 60-ft. dishes at about 300 miles. In spite of this, however, it is quite usual to base loss-of-antenna-gain calculations on the plane-wave gain without allowing for the fact that it is an effect sharply dependent on path length. In other cases attempts have been made to improve transmission over paths of moderate length by siting receiving stations at great heights above sea-level. While this nearly always has the effect of considerably increasing the average signal received, and may even to some extent lessen the maximum path loss, it will also reduce the distance between radio horizons, and may thus lead to so much antenna loss-of-gain as substantially to counterbalance any advantage attributable to the reduction in effective path length. There have been several cases in which the re-siting of receiving stations in this way has not been accompanied by the expected improvement in the minimum received signal level.

If for any reason it is desired to use antennas larger than that in which there will be, for 99% of the time, substantially a Rayleigh distribution and a single 'path', if follows from the statistical theory that the average antenna gain to be expected over roughly flat country will be less than the full 'plane-wave' gain by half the extra gain (in dB) above that of the antenna spanning the curvature range of 0·002084 curth (in SATRA); that is to say, the average loss-of-antenna gain in dB will be half the difference between the full plane wave gain and that of the antenna of which the vertical aperture dimension in feet is approximately $D^2/1{,}000$ ft., D being the path length between radio horizons in miles. The actual loss-of-gain will fluctuate over a total range from 0 dB to twice its average value.

The latitude effect

All of the figures so far given for the parameters of the turbulence statistics, for the zone height and time, and for the mixing velocities, etc., in the lower atmosphere apply to conditions averaged over the troposphere as a whole, and are based on the idealised conditions to which we have referred as SATRA. They apply therefore most accurately in dry air, at latitude $37\frac{1}{2}°$, and at the equinoxes.

The points marked in Figure 3 correspond to measured heights of the tropopause at various latitudes and at different times of the year, as given by the Meteorological Office. The full line in this figure has been drawn through points calculated in the way to which reference was made in Chapter 4, and they are so close to figures obtained from the Meteorological Office that they will be used here as the most convenient average figures available. The calculation of the height of the tropopause in Figure 3 has been based on the assumption that the turbulence is actually the result of *three* causal sets:

(a) Close to the mean latitude $37\frac{1}{2}°$, the average diurnal cycle insolation is proportional to $\cos^3\varphi$, where φ is the latitude. This is the source of the coherent vertical component in the mixing.

(b) There is at all latitudes the isotropic and equiprobable causal set associated with the solar tide and responsible for the basic temperature to which the diurnal cycle heating is additional.

(c) There will be additional turbulent mixing in the general direction from equatorial latitudes towards the Poles, with a mean direction a little below the horizontal. This is a direct consequence of the fact that the average diurnal cycle intensity is much greater nearer the Equator, and that the isotropic mixing along the layers is much more effective than the relatively slow vertical component.

If the total height of the tropopause and the change in daily insolation with latitude φ are both proportional to $\cos^3\varphi$, the latter will change almost linearly 4% per degree of latitude in the neighbourhood of latitude $37\frac{1}{2}°$. The whole of this insolation reaches the atmosphere from the ground, passing through the stratum which contains the lowest diurnal cycle of heating, and changing, in the *whole* height mixed in four days, by an average of 16% of the daily insolation per degree of latitude.

The isotropic and equiprobable perturbations at all latitudes are unchanged at 1·095% of the total. There is additional mixing along the layers, which are in fact not quite horizontal but downwards by an amount which accumulates to a total of a quarter P(R) (0·0005 curth approximately) over the whole distance from Equator to the mean latitude, and again to the same total from mean latitude to Pole. The mean flow in the zones along these layers is actually along a curvature of 1·25 curth, is equiprobable over the range from 1·0 to 1·5 curth, and occurs with a relative probability of 1·095% also. It will therefore fall through the stratum in which there is one complete diurnal cycle in a horizontal distance close to 148 miles, or 2·14° of latitude. This means of course that the volume of air all at one single 'temperature' varies a little from latitude to latitude. The total represented by both these random causal sets in the whole height of troposphere thus changes by $4\times\sqrt{2}\times 1{\cdot}095/2{\cdot}14$, or 2·9% of the daily insolation per degree of latitude at and near latitude 37½°. With increasing insolation therefore there is also this increase in the probability of the zones along any single layer temperature, and the probability (zone density) of the average coherent insolation in the neighbourhood of the mean latitude rises and falls 1·16/1·029 or 12·7% of the daily insolation per degree of latitude.

This means that the change in coherent insolation rising from the ground will degrade to random, falling or rising by 98·45% of that in one diurnal cycle, in about $\pm 98{\cdot}45/12{\cdot}7 = 7{\cdot}73^\circ$ of latitude approximately. Considered in time-intervals of about four days, changes in the average insolation which result from latitude are thus observable in the air above ground only over the latitude range of 29·7° to 45·2°. Outside this comparatively narrow range of latitude and on the average over the year, the temperature difference averaged over the whole vertical profile is more likely to be the combined result of the solar tide and the flow of heat along the layers from warmer latitudes than the diurnal insolation rising from the ground below.

The change in tropopause height outside these latitudes is, of course, predominantly the result of the flow of heat from the layers, and will be approximately 1·095% per 2·14° of latitude, that is to say, about 0·51% of the daily insolation per degree of latitude. Since inside this mean latitude range the coherent radiation coming from the ground below strongly predominates, there may well in this belt be two tropopauses, one, after four and a half days of cooling, starting from the maximum temperature during the diurnal cycle at the ground, and the other, after three and a half days of cooling, arising from the minimum night temperature at ground-level. The smooth curve of Figure 3 gives the theoretical tropopause height for different latitudes averaged over the year considered in time-intervals of about four days.

This theoretical curve seems to be usefully accurate everywhere except within the Arctic and Antarctic circles, where the actual tropopause heights are noticeably lower than the theory would suggest. There may be three reasons for this:

(a) The air in Polar regions contains very much less moisture than elsewhere, and the latent heat retained in other latitudes may result in less turbulent energy being available in Polar regions.

(b) What solar heating there is in arctic and antarctic regions as a result of the obliquity effect may to some extent be absorbed in melting ice and snow at the surface, some of this energy again passing to more temperate latitudes in latent form in the flow of water during the spring.

(c) Some of the incoming solar energy may be reflected from ice and snow, that is to say, the albedo of the earth is presumably higher near the Poles than elsewhere. Since direct insolation is very small in such areas even during summer, this is unlikely to be of importance.

The complete statistical parameters for the turbulence at all latitudes corresponding to the smooth curve in Figure 3 are given in Table 1. As already indicated the obliquity effect (annual cycle) can be estimated from these by assuming that the latitude changes by $\pm 23\frac{1}{2}°$ during the year.

Other microwave radio propagation and system characteristics

In addition to the figures already given, the statistical theory leads to much improved estimates of the diffraction losses introduced by obstacles, and the extra loss to be expected when an antenna has to be directed above the horizontal to surmount them. In particular, the theory provides a much more satisfactory explanation of the phenomenon sometimes known as 'obstacle gain'. The random field components will of course also have an obvious and important influence on the planning and effectiveness of diversity systems, and the theory can also be used to indicate the useful radio frequency band-width available in many cases.

TABLE 1: *The effect of latitude*

Latitude Degrees φ	0°	10°	20°	30°	35°	37½° SATRA	40°	45°	50°	60°	70°	80°
Height of the Tropopause (ft.) H_{tp}	54,500	53,950	53,400	52,750	47,060	43,600	40,140	35,250	34,900	34,400	33,800	33,260
Basic Uncertainty $P(R)\varphi$	0·002605	0·002579	0·002552	0·002521	0·002249	0·002084	0·001919	0·001684	0·001669	0·001643	0·001616	0·001590
Normd. Gaussn. Devn. Corresponding to $P(R)\varphi$ $U\varphi/\sigma\varphi$	2·7937	2·797	2·8004	2·8043	2·8408	2·8651	2·8912	2·932	2·9348	2·9394	2·9448	2·9498
Stnd. Devn. $\sigma\varphi = 1{\cdot}25/U\varphi$	0·4474	0·4469	0·4463	0·4458	0·4400	0·4363	0·4323	0·4264	0·4259	0·4253	0·4245	0·4238
Indeterminacy $P(R)\varphi/\sigma\varphi^2$ (100-S)%	1·301	1·291	1·281	1·266	1·161	1·095	1·027	0·9262	0·9202	0·9082	0·8966	0·8852
(100-S)% of 1 yr. Days T_1	4·75	4·72	4·68	4·625	4·24	4·0	3·75	3·39	3·36	3·32	3·275	3·23
Vert. Mixing Velocity in ft./min. $V_v = H_{tp}/T_1$	7·96	7·94	7·92	7·92	7·71	7·57	7·43	7·23	7·22	7·20	7·17	7·14
Zone Time Minutes $Z_t = P(R) \times 1{,}440$	3·75	3·71	3·67	3·63	3·24	3·0	2·76	2·43	2·40	2·37	2·33	2·29
Zone Height (ft.) $V_v \times Z_t$	29·9	29·5	29·1	28·8	25·0	22·7	20·5	17·6	17·3	17·1	16·7	16·4
Ht. Mixed in 24 hrs—ft. (Total Illum'tn.) H_{tp}/T_1	11,500	11,400	11,400	11,200	11,100	10,900	10,700	10,400	10,390	10,360	10,300	10,300
Loss to Random Field at change-over 20 log $P(R)\varphi$–dB	51·7	51·8	51·9	52·0	53·0	53·6	54·3	55·5	55·6	55·7	55·8	56·0
Loss rate to Random Field beyond change-over dB/mile	0·124	0·125	0·125	0·126	0·129	0·133	0·136	0·140	0·141	0·142	0·142	0·143

Experimental verification

It is not proposed to refer in detail to published data which support the conclusions of the statistical theory, but these are in general impressive, or highly significant, in the following areas:

(a) The mean refractive index gradient, and the shape of the distribution of other values about the mean suggested by the theory, are closely in accord with values which have been used in the United States for a number of years, though of course there is no indication in the American recommendations that the bending should be regarded as completely random whenever the deviation from the mean exceeds a limiting value.

(b) Work done at the Central Radio Propagation Laboratory at Colorado confirms that the field at some distance beyond the horizon correlates with a value of refractive index gradient which is five times as great as the mean. In the Neogaussian distribution, of course, the maximum coherent deviation is 1·25, which is five times the mean of 0·25.

(c) The fading characteristics associated with line-of-sight transmission, the signal in the diffraction region, and the statistics of the random field beyond the change-over are all closely those to be expected from the application of a Neogaussian hypothesis in dry air conditions. The presence of moisture may distort the distributions, but observed effects do not differ materially from those which might be expected by extending the theory to allow for latent heat effects.

(d) The loss at the horizon, in the diffraction section, at the change-over, and in the random or scatter field beyond change-over, agree very closely with the theoretical figures. It is believed that the statistical theory is unique in that these figures have been obtained without the use of external constants other than the assumption that the radius of the earth, and of the curvature of the layering in the atmosphere close to it, is 3,960 miles, and that the diurnal solar cycle lasts 24 hours. The fading ranges at all path lengths are approximately those to be expected from the theory. It should be remembered here that the fading range depends on the mean signal level as well as on the minimum signal level in 99% of the time. The mean signal comes, on the average, from only a quarter of the height of the atmosphere illuminated above the receiving radio horizon, and in this lowest quarter the effects of additional turbulence over rough ground and of moisture are often very marked. While, therefore, the minimum signal in 99% of the time can generally be estimated for the longer paths to within two or three dB, it is to be expected that the mean signal and fading ranges will be more indeterminate.

(e) Measured figures for the vertical mixing velocity, which controls the period of time during which substantially Rayleigh distributions are to be expected, are closely in accord with theoretical figures. Diffraction losses over obstacles can be calculated much more accurately than when classical diffraction formulae are used. The qualitative and quantitative explanations of loss of antenna gain provided by the statistical theory are much more convincing than when other explanations are accepted.

A great deal of published data cannot directly be compared with statistical deductions. In many cases signals have been averaged over periods of time such that important signal changes are often hidden in averages, and in other cases a large number of tests have been grouped together regardless of latitude differences and the moisture conditions at the time of the tests. For these reasons the statistical parameters at different latitudes and the figures derived from them should perhaps be treated with some reserve, though comparisons made over many years between published results and the theoretical deductions have not disclosed important discrepancies.

PART IV

MATTERS OF OPINION

Truth is never pure and rarely simple

8

THE PROPER STUDY OF MANKIND IS MAN

So far nearly all that has been put forward has been in support of the idea that, in many analyses of fields of changes in which natural, uncontrolled disturbances are important, rigorous mathematical assumptions are inadequate, and can with advantage be superseded by a statistical hypothesis in which there is provision for extraneous perturbations in addition to the coherent changes of admitted interest. Turbulence, and if one looks closely enough probably every other field of change in the real world, involves far more variables than can be included in any single rigorous expression, so that absolute precision of analysis is impossible. The best we can hope to do is to analyse down to the smallest identifiable coherent change, which is the one in the sub-distribution in which a change is just as likely to be due to extraneous causes as to the coherent parameters regularly included in rigorous analyses.

My argument so far has been based on statistical assumptions which, I hope, are plausible in themselves, even though they are not those generally adopted. The ease with which the assumptions of a rectangular distribution for the random perturbations alone, and a Gaussian distribution for the coherent changes alone, can be accepted is a subjective matter of which each of us must be his own judge. All that can usefully be added here is that there is a great deal of qualitative evidence, based on the actual characteristics of the sub-distributions and part-distributions which occur in natural fields of turbulence, which support these assumptions—indeed there is a great deal more such evidence than there has been space for it here.

The complete solution for the distribution from both causal sets involves two additional simplifying devices. The first of these is that the data analysed must be prepared in such a way that each observation is *always* an equiprobable (approximately rectangular) sub-distribution of perturbations, that is to say, each measurement should be of a group of perturbations during a time-interval or over an amplitude range within which the disturbances have a constant mean. The rule that turbulent fields should be analysed in sub-distributions *all* of which have the same statistical 'shape' is widely recognised, but it is also, regrettably, often ignored in practice. To obey it may necessitate measuring over intervals of time or amplitude which may be inconvenient. The other normalising device is the one which leads to a general solution which may be used in many different analyses regardless of the relative importance of the perturbations. It involves defining as unity that deviation from the mean of the Gaussian distribution at which the frequency of occurrence of a particular small coherent change is the same as that of an equal change due to the extraneous or random causal set also contributing to the total field of disturbances. Since the probability of any given change is its relative frequency of occurrence, and the incoherent causal set alone would result in a rectangular distribution, this implies that we are defining as 1·0 both the maximum coherent deviate in the distribution of changes from the Gaussian causal set, and the total range of the perturbations from the random causal set about each coherent deviate.

In Parts 2 and 3, these basic assumptions and the ensuing normalised statistical solution were applied to the analysis of four fields of natural disturbances, two of them fields of changes in the lower atmosphere, and the two others, fields of information bearing communication 'signals'. In the case of the temperature distributions in the lower atmosphere, and the gravitational field close to the earth, the normalising constants introduced are so well known that they are really beyond dispute. Perhaps the most surprising fact about these analyses is that they lead to a number of very well-known 'empirical' constants, which are thus shown to be a necessary consequence of the statistics, and are not natural constants for which no explanation can be proffered. In these analyses, too, the minimum detectable coherent changes, both in the time and in the height distributions, are larger than might have been expected, and this may account for the fact that they seem not previously to have been identified. The two analyses in Part 3, of speech and of microwave propagation through the lower atmosphere, appear to point to a most interesting connection between information theory and the statistics of the incoming changes as a result of which we see and hear—between the patterns in the changes in the world around us, and the pattern recognition activity as a result of which we can appreciate and acquire information about these changes.

The evidence in support of the statistical theory, including as it does calculations of:

(a) The height and temperature of the tropopause, the height to the base of the ionosphere, the average velocity of sound in the lower atmosphere, the mixing velocities in its turbulence, etc.;

(b) The relationship between the day and the year for maximum temperature stability, the mean orbital characteristics of the earth around the sun (for maximum orbital stability), the height of the 'stationary' satellite, etc.;

(c) The articulation content and other statistical parameters of speech, all of the more important characteristics of microwave radio propagation through the lower atmosphere at line-of-sight, diffraction, and scatter distances;

would seem to be much more extensive and significant than has been achieved by other analytical methods applied to turbulence problems. It seems also to depend on far fewer and less controversial normalising constants than other theories. I propose therefore now to assume that all of this constitutes 'proof' of the validity of the assumptions underlying the theory, and pass on to a number of deductions which, though not in this sense proved, appear to me to constitute reasonable extensions to the general hypothesis on which everything so far has been based. These deductions will be considered in three areas, one dealing with the guide-rules or laws which seem to follow from the statistics, a second referring to ways in which these rules or laws appear to suggest the need for a fundamental change of viewpoint in modern science, and a third covering one or two generalities about the relationship between man and his environment and the way in which he may be evolving as a rational being in a largely coherent world much too complex for him as yet fully to understand. These three areas together will, I hope, form the bridge between rigorous science on the one hand and the changing world on the other which it has always been my ambition to construct.

The guide-rules

1. All measuring is, in the last resort, a pattern recognition process, a statistical appraisal involving two domains. In one of these the law of the measuring scale may be predominant; in the other the statistical characteristics of the scale marks, and/or of the perturbations in what is measured, are more important.

2. If all available information is to be extracted from the data, the measuring sensitivity should be such as to reveal the limiting source of uncertainty or the perturbations, and a statistical approach used. The data should be analysed in sub-distributions all of which to a second order of approximation have similar statistical characteristics. There will then be:

(a) A basic uncertainty limiting the precision to which the analysis can be taken. This is not necessarily because of the measuring technique; it may come from turbulence within what is being measured.

(b) An indeterminacy, never less than the basic uncertainty, in all probabilities.

(c) A finite range of validity of any natural law formulated to fit the data.

3. Simply as a result of conditions in our solar system and perhaps even in a much larger astronomical system of which it is part, the important causes of natural disturbances acting from without can, to a second order of approximation, be regarded as producing Neogaussian distributions of changes. This means that in one scale of magnitudes they are coherent and form part of an approximately Gaussian distribution, but outside this scale they are apparently random and can be regarded as forming part of an approximately rectangular or equiprobable distribution. The changes thus conform to different laws with a statistical change-over between them. This is at present often overlooked.

4. Presumably as a by-product of 3, the same type of statistical analysis (application of the Neogaussian distribution) can be used to explain the pattern recognition activity by which we gain information about those fields of natural disturbances which stimulate the receptors in our eyes and ears.

The above guide-rules are in accord with the second law of thermodynamics in that they include the *tendency* to maximum randomness or entropy in natural phenomena. They refute the idea that a rigorous law can ever be true over an indefinitely wide range of natural change.

A new look at some fundamental concepts

For some years now it has been necessary simultaneously to accept scientific ideas which are, in the last resort, mutually conflicting. The wave and quantum theories in physics, the diffraction and scatter modes of radio propagation, are but two examples of this. There has also at times been a tendency to regard different parts of a 'population' each as conforming to one or other of the conflicting concepts. Communication engineers often speak and plan as if separate and distinct noise and signal currents were present together in an electrical circuit. It is surely more reasonable to assume that the whole current is simultaneously subject to two causal sets, one of which alone would produce the noise and the other the signal changes, both causal sets at all times operating on all of the current. Although it may then be impossible to decide definitely whether any particular small change is signal or noise, the probabilities of it being one or the other can be estimated, at any rate once the normalisation constants in the particular case have been found. The view that all waves are partly coherent and partly incoherent, and that small parts of them comply with the well-known laws of wave propagation to an extent determined only by a small coherent velocity component (additional to their random and isotropic movements) normal to what is, to some degree, an indeterminate wave-front, is surely much more realistic than any assumption that there is a precisely defined wave-front with completely isotropic point sources distributed along its length. At any rate the former approach does permit analysis of conditions in which the amount of turbulence and the relative level of the perturbations within small parts of the wave-front change along the path.

To take a familiar example, it is impossible for the *whole* area of a carpet to be covered by a single pattern, for then there would be no background from which any pattern could be distinguished. As soon as we admit the existence both of a pattern and of its background, we must have two separate 'laws' to express the different characteristics of each. In the case of the normalised Neogaussian distribution, the maximum fraction of any whole that can be identified as coherent pattern is 98·45%. Since all measurement is, in the last resort, a sort of pattern recognition, there must also be, associated sometimes with the measuring scales, a very small minimum probability (quantum) limiting the precision with which we can measure. This minimum is essentially a probability, the basic uncertainty always present in any information-seeking analysis. Uncertainty is not peculiar to particle physics. Rather is it an indication of a statistical change-over between domains in which different laws predominate.

Even the so-called fundamental dimensions, mass, length, and time, can be identified only within these limitations, and if the fundamental dimensions have limited spheres of validity, still more must this be the case with derivatives of them such as energy, velocity, reflection, modulation, etc. To quote an extreme example of loose terminology in this direction, it is often said that radio waves are 'reflected' from layers in space in conditions in which the matter allegedly responsible for the reflection is so sparsely distributed that it is inconceivable that layers of it can act as an effective obstacle at the wavelengths concerned. It is quite true of course that some small part of the radiation may be deflected and even returned towards earth when radio waves are transmitted some distance through such space. But surely it is more reasonable to refer to this as the result of the turbulence in large volumes rather than reflection from discrete layers.

Perhaps the clearest indications of the important difference which a statistical outlook can introduce, can be seen by comparing some of the fundamental ideas in the Theory of Relativity with the corresponding statistical concepts. The statements in the left-hand column below have been taken (not always verbatim) from the earlier pages of Albert Einstein's *Relativity: The Special and the General Theory—A Popular Exposition*. In the right-hand column are similar ideas more in consonance with the statistical view-point.

Geometrical ideas correspond to more or less exact objects in nature and these last are undoubtedly the exclusive cause of the genesis of those ideas.	There are in nature very few objects which can be expressed even approximately in geometrical terms. The approximations involved are nearly always very much larger than the relativistic corrections for the velocity of propagation of light-waves.
There is no such thing as an independently existing trajectory, but only a trajectory relative to a particular body of reference.	Statistically a trajectory is a description of a succession of positions no two of which are occupied at any one time. No trajectory ever exists as a whole, therefore; in the last resort the line selected to represent it depends not only on the body of reference but also on the part, size, and

shape of the body considered to be traversing it. The trajectory is in fact a synthesis, of the coherent elements only, in both the time and space distributions, making up the path, all perturbations (or other relative movements of parts such as are involved in rotation) being ignored.

We call this motion a uniform translation ('uniform' because it is of constant velocity and direction, 'translation' because although the carriage changes its position relative to the embankment yet it does not rotate in so doing).

Uniform translation is a theoretical concept of motion devoid of all perturbations.

If, relative to K, K′ is a uniformly moving coordinate system devoid of rotation, then natural phenomena run their course with respect to K′ according to exactly the same general laws as with respect to K.

Natural phenomena run their course approximately in accord with general laws of limited validity, which must be changed whenever we take into consideration occurrences within sufficiently smaller or sufficiently larger time-intervals and/or distances, such as those generally involved when correcting for the finite velocity of light-waves.

There is hardly a simpler law in physics than that according to which light is propagated in empty space. Every child at school knows, or believes he knows, that this propagation takes place in straight lines with a velocity $C = 300{,}000$ km./sec.

Space containing light is not strictly empty. The behaviour of light, as judged by its wave structure, must depend on the conditions at its source and changes in its condition during transmission from source to the space considered. Since light has so many of the characteristics of extremely sparsely distributed matter, the movements of one part of a light-field will presumably be to some extent affected by other parts in proximity with it. Any coherence at source will, in the absence of all other factors, tend to persist. References to the velocity of light seem really to refer to the velocity of light-*waves*, and it is likely that this latter will depend on the statistics of the gravitational field, including perturbations, acting on the light during its transmission to and through the so-called empty space.

If a ray of light be sent along the embankment.

A 'ray' of light appears to refer to a small amount of radiation embodying the approximate large-scale coherent

> characteristics of light-waves—constant velocity in a straight line, at right-angles to a coherent wave-front. The concept of a ray is apparently the result of a desire to express these large-scale coherent characteristics without introducing the perturbations which in fact are always to be found within a complete wave. The full behaviour of light-waves can be explained only by assuming that very small parts of all waves act as isotropic sources, and very little in ways representative of a wave as a whole. If this is the case, then a ray of light may be a useful concept when the coherent behaviour of complete waves is to be considered, but rays do not really exist. They are concepts which have only the limited sphere of validity of the waves themselves.

Similar comments could no doubt be applied to the rest of this book. I do not propose to do so, however, both because I regard the work as one of genius and because more such comment would be largely unconstructive. I prefer to assume that, perhaps to some extent because it is a 'popular' exposition, the author has failed to point out that, in it:

(a) 'Light' is discussed only in terms of its coherent components, that is to say, those which predominate in its waves, and not those which, in the main, produce the 'isotropic sources' in small parts of the wave-fronts.

(b) By 'velocity of light' is meant the velocity of light-waves. Most practical measurements of the velocity of light consist of measurement of the straight line distance along which light-waves are presumed to pass, and of the time taken for the light-waves to traverse this path. It is of course well known that an intermediate opaque obstacle, small compared with the wavelength, does not cast a 'black' shadow, so that, *considered in very small parts of a wave-length*, the light does not travel in straight lines. As far as I know, no one has as yet succeeded in measuring any 'velocity of light' other than that of the coherent components contributing to its waves.

(c) The explanations given by Einstein for the Lorentz transformations may be replaced by considerations attributing the pattern distortions in length, mass, and time involved in these transformations to perturbations. If this is done it will be seen that they are valid approximations at velocities less than that of light-waves in vacuum for particles (rigid bodies) the disturbance of which leads to the emanation of waves of radiation. In the limiting condition, when the velocity of a rigid body approaches that of light, the concepts of a rigid body and of coherent waves fail, since the body would presumably be in a state of transition to radiation; at these velocities the coherent forces holding it together have presumably degraded to the random perturbations always to be associated with them. If this is true, matter moving with the velocity of light-waves can exist in *stable* form only as radiation. Similarly, if we imagine a reference scale travelling with the velocity of light-waves, there will remain relative to this scale within the waves only those random velocity components appropriate to the isotropic point sources along the wave-front; as seen from such a measuring scale, the coherence essential for us to observe waves can no longer be identified.

The transfer of the smallest particles of matter from one stable orbit to another involves a readjustment of the total energy distribution between that in the perturbations and that of the complete particle in orbit, in (to a second order of approximation) discrete quanta of energy. This is presumably because, by definition, a stable orbit is one in which the perturbations, rotations, or other relative movements within the particle do not accumulate sufficiently to disturb the stability of the orbit within the lifetime of the particle. It would seem again therefore that the appearance of quanta is an indication that the statistics of the perturbations (or spin) must be taken into consideration in any complete explanation of the phenomena. More than this, at the velocity of light-waves and in otherwise empty space, it is apparently impossible by direct observation to distinguish between small differences occurring in the space distributions at any one time and small changes with time occurring at any one place. When the coherent components in both time and space distributions degrade to random, neither can be identified. This is the limit to our innate pattern recognition activity, the basic uncertainty in the statistical appraisals as a result of which we extract information from the fields of changes stimulating the sensitive nerves in our eyes and ears. It might be possible of course to *infer* what is happening, at any rate to a second order of approximation. It seems, however, that as far as we can tell at present, direct observation of coherent changes in bodies moving at speeds in the neighbourhood of the velocity of light-waves, is beyond us.

The English translation of Einstein's book was first published as long ago as 1920. Nearly forty years later there appeared *The Physicist's Conception of Nature*, by Werner Heisenberg. The philosophy underlying this work seems to be in much closer accord with the statistical approach I have tried, no doubt much less successfully, to put forward here. To quote but three statements from this work:

(a) 'One needs conclude that all natural laws may be considered to be natural statistical laws.

(b) 'Quantum theory actually forces us to formulate these laws precisely as statistical laws.

(c) 'In very small regions of space–time, of the order of magnitude of the elementary particles, the notions of space and time become unclear.'

The only comment I can make on these three quotations is that I would have omitted the word 'precisely' in (b) and replaced the word 'unclear' by 'uncertain' in (c).

Man and his environment

It is of course well known that natural changes, without human interference, tend toward a condition of maximum randomness (entropy). Man's activities, on the other hand, are as a rule directed toward reducing extraneous perturbations so that coherent patterns can be more easily recognised, and events anticipated and, sometimes, even controlled. We are ourselves part of nature, and our senses are remarkably efficient at acquiring information about changes in our environment from the patterns we can recognise in them, using our eyes and ears. The recent rapid rise in the use of machines in the home and in industrialisation of our societies has brought with it a great increase in the amount of measurement, and the development of an enormous range of instruments of remarkable sensitivity and accuracy for this purpose. But although the precision with which we can measure objects and events is very great indeed, there are always fundamental limitations to measuring accuracy, just as there are in other forms of pattern recognition. These limitations may prevent us from taking full advantage of the extreme sensitivity of our instruments in the following ways:

> Ultimately, when the time-interval involved in a change is so short that, within it, no coherent pattern (wave) can build up in the light reaching our eyes, we can neither time nor identify the change. This is, I believe, the 'relativistic' limit to our powers of observation.
>
> The effects of the measuring technique itself on what we are trying to measure may result in changes greater than instrumental errors. This is, I believe, what is often called the Heisenberg Uncertainty Principle.
>
> There are always perturbations in the measuring scale as well as in what is measured, and these may introduce indeterminacy greater than the average errors of measurement. There are no perfectly rigid bodies, and no precisely single-valued parameters, and if the perturbations in them are statistically important enough, it may prove impossible to make consistent measurements, let alone formulate coherent laws to explain them.
>
> Perturbations in what is measured, in time and/or space, may extend over such distances and time-intervals that all ordinary measurements fall below the change-over point at which coherent patterns start to appear, so that successive measurements may bear no discernible consistent relationship one to another. If such a field of changes continues for some time, it is generally referred to as turbulence. In other cases it may be classified as a 'transient' phenomenon.

All of the above limitations are essentially of the same type. They are statistical, and whenever they apply any attempt at a rigorous analysis which neglects extraneous perturbations must clearly fail. Mathematics is as yet mostly a coherent language, and even in Statistics it is very often assumed that there are precise probabilities and laws which apply over indefinitely large ranges of deviation. The Neogaussian distribution includes random perturbations, and for this reason is often a most useful approximation to conditions in natural turbulent fields, complete description of which would involve an indefinitely large number of coherent laws, and statistics of a complexity far beyond our present powers.

Its normalised solution appears to provide a remarkably good fit to changes both in the natural world of which we are part and in the sensibilities through which we observe the world. As it includes a change-over between two domains each on a different scale, it is sometimes possible to apply it two or three times in succession, in a gradual approach to the whole, probably unattainable, truth.

Sensibility

We can perhaps gain a little insight into the way in which we acquire information about our environment from the statistics to which our senses of sight and hearing conform. The behaviour which seems best to fit the facts is one in which our eyes and ears slowly and continuously 'adapt' their threshold sensitivity to a background to which, as a rule, we pay little attention. Disturbances which are large enough overflow from one stimulated receptor to the next. If as a result of a coherent relationship between successive groups of perturbations there is a temporary larger-scale change at the start of some recognisable pattern, the extent to which this change contributes information can be measured by the logarithm of the relative number of adjacent nerves activated in any group contributing to a coherent pattern, on a scale on which the threshold sensitivity, to which the rest of the array of nerves has adapted, is zero, Since the background itself is not steady but a random (incoherent) field of perturbations, such information is essentially a probability with a background indeterminacy and still longer-term basic uncertainty.

We cannot of course be continuously aware of each separate stimulation in every single receptor in all our senses. There must be a highly restrictive selective mechanism somewhere in the mind, accepting only those changes which contribute to a coherent pattern. Here I propose to refer to the combination of each sense with this selective mechanism as 'sensibility'.

All the evidence I have so far proffered appears to support some such hypothesis as the above, for our sight and hearing. It seems likely, however, that our whole sensibility, and not just that part of it associated with these two senses, is organised for pattern recognition. For this to be so, there would need to be innate in us:

A subconscious ability continuously to adapt all our threshold sensitivities to the mean of their background. In Neogaussian statistics this corresponds to adjustment to zero coherent deviation in the distribution, where sensibility is a maximum, though still limited, of course, by the background perturbations.

An urge to search for any pattern simple enough to identify and perhaps memorise in some way. This implies a sensibility continuously seeking the coherent components among the incoming changes and generally sensitive enough to recognise them right down to the perturbations.

These together would clearly allow us to recognise vestiges of a familiar pattern even when the distribution in by far the greater part of the array of stimulated receptors is apparently patternless. The cocktail-party effect, and the ability to decipher almost illegible hand-writing, are well-known examples of this.

Let us consider the course of human evolution to be expected assuming such innate mental capabilities. As by far the greater part of the earth's surface is covered by water, it seems most likely that life originated in the seas. But the perturbation levels obscuring the coherence in the changes from the two most important external causal sets of natural changes (temperature and gravitation) are extremely high in this environment, so that a form of life with an urge to search for coherent patterns would tend to migrate to dry land. Once there, the next step would be presumably to adopt an erect posture in which the larger concentrations of receptors (eyes and ears) are raised clear of most surrounding obstruction, the body as a whole remaining within the 'single-temperature' lowest zone (about 22·7 ft. of height) of maximum diurnal cycle heating. Within small parts of this zone coherent temperature *changes* are not statistically detectable, but there is, in the *gravitational* perturbations associated with the solar tide, a coherent diurnal pattern with its mean, that developed in one-quarter of a day, extending over a height intercept of about a quarter of 22·7 ft., or 5 ft. 8 in. above ground. A body placed to extend vertically above ground for about 5 ft. 8 in., with its centres of maximum concentration of receptors near the top, would be in the position of maximum advantage from the point of view of pattern recognition.

Whether this is the right explanation of what actually seems to have occurred is, no doubt, a matter of opinion. It appears to be significant, and it may be that the advantage thus gained lies in the improved sense of balance which might come from extreme sensibility to coherent pressure change, and in the ability it might give to time events through their relation to the diurnal gravitational (solar tide) cycle, and thus enable us to regulate the cyclic controls built into many of our bodily functions.

Evidence of the existence of these same tendencies in our social evolution may be seen when we consider the origins of our longer lasting and more highly developed religions and civilisations, though in this case they can apply only so far as natural conditions facilitate communication between individuals, and hence the development of coherent communities. In this connection, we will take as evidence of coherence the development of:

Planned Agriculture (under the control of the annual cycle of diurnal insolations?)
Urbanisation
Formalised systems of Religion, Morality, and Law
Spoken and Written Language
Tools and Machines
Libraries and other information stores
The Arts and Sciences
Transport and Telecommunications.

We might then expect such coherent communities (civilisations?) to have developed *first*

(a) In those areas in which the diurnal cycle insolation from the ground is, even in the presence of extraneous perturbations from other latitudes, predominant. In Chapter 7, page 126, these were shown to be the land areas between latitudes 29·7° and 45·2°, i.e. those above which there are very often two tropopauses.

(b) Especially in such areas where there are good natural communications, i.e. round inland seas, along navigable rivers, etc.

(c) Occasionally in other latitudes at high altitudes where mean temperatures are low enough to permit recognition of the coherent annual temperature cycle in the lower perturbation levels associated with the lower temperatures.

While our knowledge of the origins of the more important religions and civilisations is not extensive, what is known seems strongly to suggest that pattern recognition has been of paramount importance in the evolution of man both as a rational individual and as part of a coherent society. There is, of course, almost unlimited evidence of the presence of such tendencies within ourselves today—the creation of patterns in speech and writing to call to mind classes of objects and events, qualities and abstract ideas of all kinds, the satisfaction we derive from simple relationships (harmonies) in the arts, and in our games and toys, are obvious examples. In science and technology, more often than not we first mentally create a pattern incorporating as clearly as possible the features of interest in any object or event under study, and then measure this pattern to acquire information about it. Little wonder that there are inconsistencies between the reality and the elementary patterns we so frequently bring to mind in its place.

One final 'matter of opinion' would seem not to be out of place here. However close may be the link between man and his environment, between his bodily structure and that of other forms of life on earth, in this all-important pattern activity man seems to be on a different plane from all other living creatures. It is one thing to imagine life responding to the stimulations received from its immediate environment. Machines may one day do just this. It is quite another to understand how man, without external aid, can adapt, not to the actualities which surround him, but to those patterns in them which are of special importance to his long-term evolution.

There are many forms of life on earth which are, and have been, apparently as well placed as man to acquire the information about our environment and the power over it which we now have. The facts that only man has done so, and that he alone has the conscious ability to create the patterns without which much of this knowledge and power cannot be acquired, point to some fundamental difference between him and the rest. Physically we may be a natural development from 'lower' forms of life. Mentally, man seems to be much more like that basic contradiction in terms, a fully self-programming computer. Nearly all the evidence in support of pattern recognition as the mainspring of a continuous mental evolution applies to man and man alone. In the various distributions of life, as in other fields, there are many obvious points of change-over from one domain to another—from vegetable to animal, from animal to man, and many others. They seem always to have been from lesser to greater coherence, in a direction the reverse of that in the material world, so that today, with his conscious creation of patterns which both approximate to the real world and conform to his own mental limitations (again to quote Heisenberg), 'Man for the first time now confronts himself'. He should do so, I suggest, not in the belief that there are precise, purely objective answers to his endless questioning, but with an awareness that the information he acquires stems both from within and without, and that it must be, for that very reason, basically statistical and, in the end, uncertain.

INDEX

'. . . and since perturbations are always involved, most of the terms will be used with their ordinary dictionary or encyclopaedia meanings, and not with the more precise meanings given them in technical dictionaries, in which it is often assumed that the "extraneous" disturbances of special interest here do not exist.'

Illustrations are noted in bold type